T0336811

Power Electronics Applied to Industrial Systems and Transports 3

Series Editor
Bernard Multon

Power Electronics Applied to Industrial Systems and Transports

Volume 3
Switching Power Supplies

Nicolas Patin

First published 2015 in Great Britain and the United States by ISTE Press Ltd and Elsevier Ltd

ISTE Press Ltd
27-37 St George's Road
London SW19 4EU
UK

www.iste.co.uk

Elsevier Ltd
The Boulevard, Langford Lane
Kidlington, Oxford, OX5 1GB
UK

www.elsevier.com

For information on all Elsevier publications visit our website at
http://store.elsevier.com/

British Library Cataloguing in Publication Data
A CIP record for this book is available from the British Library
Library of Congress Cataloging in Publication Data
A catalog record for this book is available from the Library of Congress
ISBN 978-1-78548-002-7

Printed and bound in the UK and US

Contents

PREFACE ix

CHAPTER 1. NON-ISOLATED SWITCH-MODE POWER SUPPLIES 1

1.1. Buck converters 1
1.2. Dimensioning a ferrite core inductance 5
1.3. Boost converters 7
1.4. Buck–boost converters 9

CHAPTER 2. ISOLATED CONVERTERS 19

2.1. Forward converters 19
2.2. Flyback converters 24
2.3. Dimensioning a flyback transformer 28
2.4. Dimensioning a forward transformer 33
2.5. Snubbers 35
2.5.1. Impact of transformer leakage inductance in a converter 36
2.5.2. Implementation and dimensioning of a snubber 37

CHAPTER 3. RESONANT CONVERTERS AND SOFT SWITCHING 41

3.1. Soft switching 41
3.1.1. Definitions, ZVS and ZCS switching 41
3.1.2. Resonance 42
3.2. Study of a resonant inverter 43
3.2.1. Presentation 43
3.2.2. Operating model 44
3.2.3. Impact of the operating frequency 45
3.2.4. Power behaviors at variable frequency 47
3.3. Study of the full converter 48
3.3.1. Analysis of the diode rectifier 48
3.3.2. Characteristics and control modes 51
3.3.3. Application to contactless power supplies . . 57

CHAPTER 4. CONVERTER MODELING FOR CONTROL 59

4.1. Principles 59
4.2. Continuous conduction modeling 60
4.2.1. The buck converter 61
4.2.2. The buck–boost converter 64
4.2.3. The boost converter 67
4.3. Discontinuous conduction modeling 67
4.4. PWM control modeling and global modeling for control 67
4.5. General block diagram of a voltage-regulated power supply 69

CHAPTER 5. CASE STUDY – THE FLYBACK POWER SUPPLY 71

5.1. Specification 71
5.2. Dimensioning switches 72
5.3. Calculation of passive components 76
5.3.1. Output capacitors 76
5.3.2. Coupled inductances 82
5.4. Dimensioning coupled inductances 83

5.4.1. Choice of a ferrite core 84
5.4.2. Windings 89
5.4.3. Tests and leakage measurements 92
5.5. Transistor control and snubber calculation . . . 94
5.5.1. Determining gate resistance 94
5.5.2. RCD snubber circuit 96
5.6. PWM control and regulation 96
5.6.1. PWM controller 96
5.6.2. Galvanic isolation of controls 97
5.6.3. Notes on modeling and control 99
5.6.4. Regulator tuning 100
5.6.5. Production 101
5.6.6. Simulations and experimental results 102

APPENDIX 1 111

APPENDIX 2 131

BIBLIOGRAPHY 159

INDEX 167

Preface

Volume 3 of this series deals with a specific category of converters for power electronics in the form of switch-mode power supplies. The main function of these components is to provide a continuous voltage to a load, smoothed by filtering elements (and sometimes regulated). In many cases, this requires a high-quality voltage supply (e.g. for electronic chips including components such as microprocessors) to guarantee successful operation. In this context, linear power supplies (ballast type) are often used, but switch-mode supplies are increasingly widespread, allowing high efficiency, and potentially improving the battery life of mobile equipment, for instance, alongside a reduction in the size of cooling elements and/or component heating.

Two main families of switch-mode power supplies will be considered in this volume: non-isolated power supplies (*buck*, *boost* and *buck–boost*) will be covered in Chapter 1, while Chapter 2 will cover isolated power supplies (*flyback* and *forward*). The list of structures presented does not provide exhaustive coverage of topologies found in publications on the subject, but it covers most requirements and includes most of the solutions used in industrial contexts. Note, however, that all of these converters operate using "hard" switching (i.e. with significant switching losses). This means that frequency increases are difficult, or impossible, preventing the miniaturization of passive components used in filtering. To overcome this difficulty, some converter topologies use the resonant behavior of "LC"-type cells, generating a considerable reduction in losses by carrying out "soft" zero voltage switching (ZVS) or zero current switching (ZCS). These converters will be presented in Chapter 3, based on a structure using a resonant inverter associated with a rectifier. The regulation issue mentioned above belongs to the field of automatics (and lies outside the scope of this book), but proportional integral (PI) regulator tuning approaches (for closed-loop control) generally pass through a modeling stage. Chapter 4 presents Middlebrook's approach for average modeling of switch-mode power supplies, which is used to define converter transfer functions. This method will be applied to non-isolated power supplies in order to establish models for continuous conduction mode. We will simply describe the models used in discontinuous mode, which are more difficult to obtain. In conclusion to this volume, we will present a case study of the detailed design and dimensioning of a *flyback* power supply, including the choice of power components (e.g. transistor, diodes, coupled inductances and capacitors) and control elements (e.g. metal oxide semiconductor field effect transistor (MOSFET) gate driver, isolated voltage and current measurements) connected to a microcontroller.

This volume also includes two appendices. Appendix 1 provides general formulas for electrical engineering (and is identical to that included in previous volumes). Appendix 2 supplies the full data sheets for key components of the flyback power supply studied in Chapter 5.

Nicolas PATIN
Compiègne, France
February 2015

1

Non-Isolated Switch-Mode Power Supplies

1.1. Buck converters

The buck converter is a single-quadrant chopper, as studied in Chapter 1 of Volume 2 [PAT 15b]. The "load" is made up of an inductance L in series with the association of the actual load (presumed to be a current source I_s) in parallel with a filtering capacitor C (see Figure 1.1).

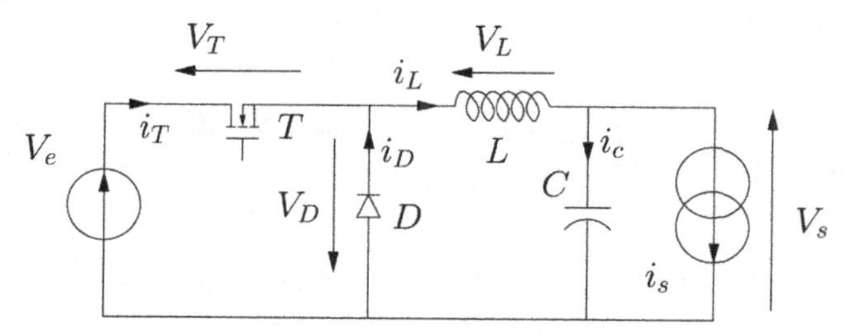

Figure 1.1. *Diagram of a buck converter*

In these conditions, for a correctly dimensioned power supply, the assembly (I_s, C) may be considered to be analogous to the electromotive force (e.m.f.) E_a of a direct current (DC) machine, and the inductance L may be considered to play the same role as the armature inductance in the machine. Consequently, the results established in

Chapter 1 of Volume 2 [PAT 15b] are applicable here. In the case of continuous conduction, an output voltage of $V_s = \alpha.V_e$ is obtained, where α is the duty ratio of the transistor control. Moreover, in cases of discontinuous conduction (i.e. for a current which cancels out in the inductance), the output voltage will be higher than in the continuous conduction case, in accordance with the characteristic shown in Figure 1.4 (see Chapter 1 of Volume 2 [PAT 15b]). A summary of the characteristics of this converter is shown in 1.1. The constraints applicable to the switches are similar to those for a one-quadrant chopper powering a DC machine, but we should also analyze the quality of the voltage supplied to the load. The waveforms produced are the same as those shown in Figure 1.2 (Volume 2, Chapter 1 [PAT 15b]), as the ripple of the output voltage V_s is considered to be a second-order phenomenon, negligible when calculating the ripple of current i_L in inductance L (constant V_s, as for the e.m.f. E_a of a DC machine). Thus, this current may be considered (as in the case of a machine power supply) to be a time-continuous, piecewise-affine function, which may be written (presuming that the load current i_s is constant and equal to I_s) as:

$$i_L(t) = I_s + \widetilde{i}_L(t) \quad [1.1]$$

where $\widetilde{i}_L(t)$ is a signal with an average value of zero, with a "peak-to-peak" ripple Δi_L expressed as:

$$\Delta i_L = \frac{\alpha.(1-\alpha).V_e}{L.F_d} \quad [1.2]$$

where $F_d = 1/T_d$ is the switching frequency and α the duty ratio of the control of transistor T.

Second, given the current ripple $\widetilde{i}_L(t)$, the (low) ripple of the output voltage $v_s(t)$ may be deduced, insofar as:

$$v_s(t) = V_s + \widetilde{v}_s(t) \quad [1.3]$$

with:

$$V_s = \alpha . V_e \qquad [1.4]$$

and:

$$\widetilde{v}_s(t) = \widetilde{v}_s(t_0) + \frac{1}{C}\int_{t_0}^{t_0+t} i_C(\tau).d\tau = \widetilde{v}_s(t_0) + \frac{1}{C}\int_{t_0}^{t_0+t} \widetilde{i}_L(\tau).d\tau$$

[1.5]

Using this result, it is then easy to deduce the peak-to-peak ripple Δv_s of voltage $v_s(t)$:

$$\Delta v_s = \frac{1}{C.}\cdot\frac{1}{2}\cdot\frac{T_d}{2}\cdot\frac{\Delta i_L}{2} = \frac{\alpha.(1-\alpha).V_e}{8L.C.F_d^2} \qquad [1.6]$$

These results are illustrated in Figure 1.2.

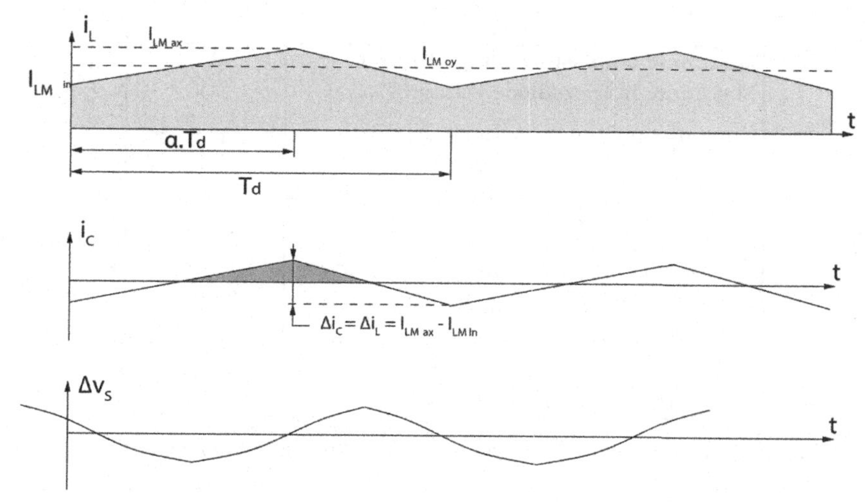

Piecewise parabolic evolution of the output voltage

Figure 1.2. *Current in the inductance and voltage at the capacitor terminals in a buck converter*

REMARK 1.1.– In practice, it is important to dimension the capacitor correctly so that the ripple of the output voltage is low in relation to its average value (e.g. 1%) in order to guarantee the validity of the reasoning used above. The current ripple was calculated based on the assumption that the output voltage is constant; strictly speaking, this assumption is not fulfilled, but the simplification is verified in practice. The decoupling of inductance and capacitor dimensioning is widespread when designing switch-mode power supplies, and will be used again when studying other structures. While this reasoning approach may appear artificial, it is based on "auto-coherence" between the initial hypotheses and the desired dimensioning objective. In practice, the output voltage should be as constant as possible when powering electrical equipment using switch-mode supplies (DC/DC converters).

Quantities	*Values*
Maximum transistor voltage V_{Tmax}	V_e
Maximum inverse diode voltage V_{dmax}	$-V_e$
Current ripple in inductance Δi_L	$\frac{\alpha.(1-\alpha).V_e}{L.F_d}$
Average output voltage $\langle V_s \rangle$	$\alpha.V_e$
Output voltage ripple Δv_s	$\frac{\alpha.(1-\alpha).V_e}{8L.C.F_d^2}$
Maximum current in transistor and diode	$I_s + \frac{\Delta I_L}{2}$
Average current in transistor $\langle I_T \rangle$	$\alpha.I_s$
Average current in diode $\langle I_d \rangle$	$(1-\alpha).I_s$
RMS current in transistor (for $\Delta i_L \ll I_s$)	$\sqrt{\alpha}.I_s$
RMS current in diode (for $\Delta i_L \ll I_s$)	$\sqrt{1-\alpha}.I_s$

Table 1.1. *Summary of continuous conduction in the buck converter*

Note that the calculation of the output voltage ripple (used in dimensioning the filter capacitor C) corresponds to a continuous mode of operation. This is not strictly applicable for discontinuous mode, but this choice presents certain advantages:

– simpler calculations using this operating mode;

– the results obtained allow satisfactory dimensioning of capacitors (on the condition that a minimum safety margin is respected) for all possible cases;

– continuous conduction is often the most critical case with regard to the output ripple, although this is not true for buck converters. The output ripple is proportional to the charge current (for boost and buck–boost converters, described in the following sections), and is higher in continuous mode (discontinuous conduction = low charge).

1.2. Dimensioning a ferrite core inductance

For iron core windings (as discussed in Chapter 5 of Volume 1 [PAT 15a]), a simple magnetic circuit was considered, characterized on the sole basis of three geometric parameters (reduced to two parameters) which needed to be established. While the equation model of the inductance and the applicable usage constraints remain identical, the geometry of a ferrite core is required, and a core must simply be selected from the lists supplied in manufacturer catalogs. The first stage in this process is to choose a family of ferrite cores: this choice depends on the application and the available space. Note the existence of "E,I" structures (alongside double E structures); in power electronics, however, the RM and PM families are most interesting in terms of electromagnetic compatibility, as they are relatively "closed" and produce limited radiation into the immediate environment. Two examples of these families of cores are shown in Figure 1.3.

Once a family of cores has been selected, we need to choose a specific model in accordance with a given specification. To do this, the expressions of A_e (iron section) and S_b (windable section) are used. Note that two surfaces are linked to constraints relating to ferrite (magnetic flux density B_{max}) and copper (current density J_{max}):

$$A_e = \frac{L.I_{max}}{n.B_{max}} \quad [1.7]$$

and:

$$S_b = \frac{n.I_{RMS}}{K_b.J_{max}} \quad [1.8]$$

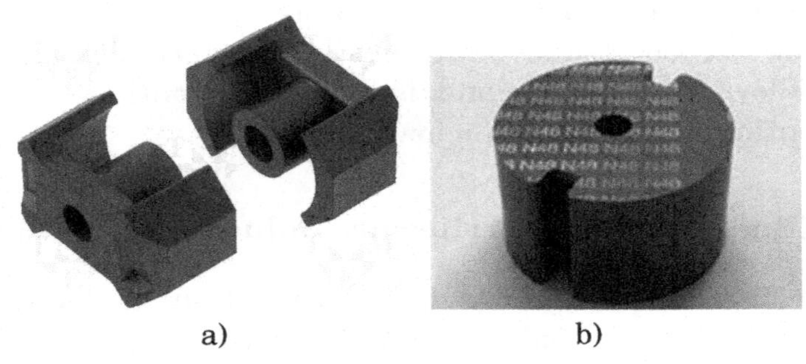

a) b)

Figure 1.3. *RM a) and PM b) ferrite cores*

The obtained product $A_e.S_b$ must be lower than that of the selected core in order to establish an acceptable solution.

Once these geometric elements have been fully designed, we may design an inductance, defining an air gap and calculating the number of turns required in the winding. Note that matching ferrite cores is particularly difficult, and it is therefore best to select a core which already contains an air gap in the central leg. In this case, the manufacturer gives a parameter known as the specific inductance, conventionally denoted as a_L (homogeneous to an inductance). This parameter is simply the inverse of the reluctance of the obtained magnetic circuit. Hence:

$$L = \frac{n^2}{\mathcal{R}_{circ}} = a_L.n^2 \quad [1.9]$$

in order to calculate the number of turns required.

Note that the inductance B_{max} required for a ferrite core is considerably lower than that used for an FeSi core (of the order of $0.2\,\text{T}$, in practice). This value may be determined by consulting manufacturer tables, which indicate losses in the material as a function of B_{max} and of the operating frequency. This loss constraint needs to be respected, as ferrites are subject to thermal runaway: as losses increase, the properties of the material are degraded, increasing the losses generated at a given operating point (leading to further temperature increases, until the limit at which the material loses its magnetic properties is reached). Note, however, that the main advantage of ferrites resides in their capacity to operate at high frequencies (in comparison with FeSi cores), as the materials used are highly resistive (and can be assimilated to electrical insulators).

While inductances in switch-mode power supplies (buck or forward type) are used to smooth current ripples at high frequencies, the skin effect remains limited, as the current ripple generally only represents between 10% and 20% of the average current; this is generally negligible, and does not justify the use of Litz wires, as in flyback and forward isolated power supplies, which will be studied in the following section.

1.3. Boost converters

Boost converters are obtained using a parallel chopper, replacing the machine (in braking mode) by a continuous voltage source V_e in series with an inductance, and replacing the voltage source E by a parallel R, C-type load, as illustrated in Figure 1.4.

The characteristics of this converter are shown in Table 1.2. Note the inversion of the problem: we no longer require the output voltage used for machine braking, but rather an input voltage V_e (equivalent to a machine being driven at constant speed); the converter output voltage V_s may therefore fluctuate. As in the case of the buck converter, the expressions of the average and root-mean-square (RMS) currents given for the transistor and the diode are based on the presumption of low current ripple in the inductance ($\Delta i_L \ll \langle I_L \rangle$). Furthermore, the output characteristic is fundamentally different from that obtained in the case of a machine providing energy to a constant voltage source: the output voltage V_s is always greater than the input voltage E, and this voltage itself tends toward infinity in discontinuous conduction in the case of a zero-loss converter (including the inductance L and the capacitor C). As an exercise, the same method to study this kind of operating mode can be applied with the buck-boost converter presented in the following section.

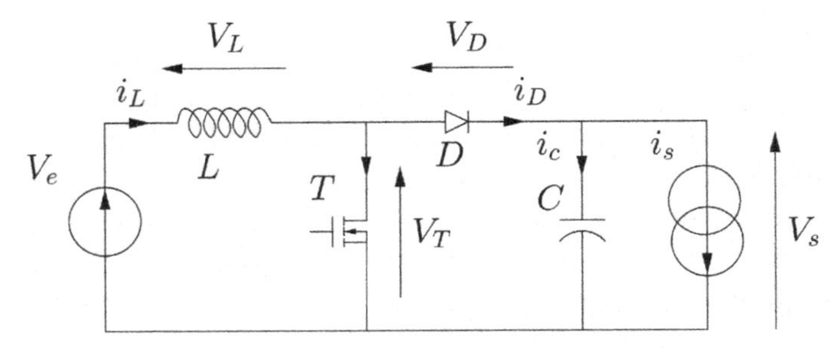

Figure 1.4. *Diagram of a boost converter*

Considering the ripple of the output voltage ΔV_s, we can easily show that the voltage drops off when the transistor is on (between 0 and $\alpha.T_d$). The capacitor therefore presents a continuous current discharge, giving us an expression of ΔV_s of the form:

$$\Delta V_s = \frac{\alpha.I_s}{C.F_d} \qquad [1.10]$$

All of the characteristic quantities of the converter (used for dimensioning) are presented in Table 1.2.

Quantities	*Values*
Maximum transistor voltage V_{Tmax}	V_s
Maximum inverse diode voltage V_{dmax}	$-V_s$
Current ripple in inductance Δi_L	$\frac{\alpha.V_e}{L.F_d}$
Output voltage ripple Δv_s	$\frac{\alpha.I_s}{C.F_d}$
Average voltage at load terminals V_s	$\frac{V_e}{1-\alpha}$
Maximum current in transistor and diode	$I_s + \frac{\Delta i_L}{2}$
Average current in transistor $\langle I_T \rangle$	$\frac{\alpha.I_s}{1-\alpha}$
Average current in diode $\langle I_d \rangle$	I_s
[1 mm] RMS current in transistor (for $\Delta i_L \ll \langle i_L \rangle$)	$\frac{I_s.\sqrt{\alpha}}{1-\alpha}$
RMS current in diode (for $\Delta i_L \ll \langle i_L \rangle$)	$\frac{I_s.\sqrt{1-\alpha}}{1-\alpha}$

Table 1.2. *Summary of continuous conduction in the boost converter*

1.4. Buck–boost converters

Buck–boost converters combine the voltage-reducing capacity of the buck converter with the voltage-increasing capacity of the boost converter. These two converters may be associated in a cascade layout to produce the same result, but it is better to combine these elements into a single structure, with a single transistor T and a single diode D, as illustrated in Figure 1.5. This converter type is also known as an inductive storage chopper, due to the way in which it operates, as demonstrated by the calculations below.

Considering transistor control strategies over the time interval $[0, \alpha T_d]$, take:

$$V_T = 0 \quad [1.11]$$

hence:

$$V_L = V_e \quad [1.12]$$

and thus:

$$V_D = -V_s - V_e < 0 \tag{1.13}$$

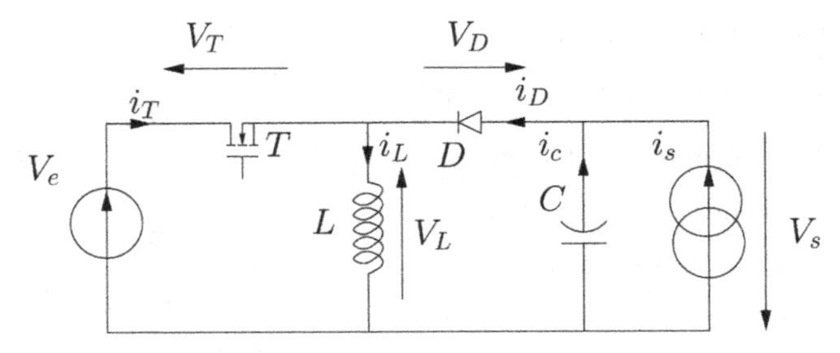

Figure 1.5. *Diagram of a buck–boost converter*

It is already evident that this negative voltage causes the diode to switch off, hence:

$$i_D = 0 \tag{1.14}$$

and:

$$i_T = i_L \tag{1.15}$$

and, additionally:

$$i_s = -i_c \tag{1.16}$$

Capacitor C therefore discharges from the output to the load. In this context, the capacitor clearly operates as an energy reservoir when the input is disconnected from the output. This initial phase of the switching period involves the storage of magnetic energy from the input in inductance L, while the load takes energy from the capacitor C. This structure operates in two distinct phases: "storage" and "restitution" of the energy stored in the inductance; clearly, it is best to obtain full demagnetization of the inductance at the end of the switching period, as stored, non-released energy is useless and will lead to overdimensioning of the inductance. Discontinuous conduction, involving an "inert" phase in which the switches (transistor and diode) are both off, is symptomatic of poor component use, and leads to overdimensioning of the decoupling capacitor in order to conform to specifications imposing a maximum ripple in the output voltage. Based on this observation, we may presume that optimal operation for this type of converter occurs at the limit between continuous and discontinuous modes, which is known as the critical conduction point. In this case, it is easy to establish the equation of current i_L, since (as in the case of discontinuous conduction) we have zero current in the inductance at the start of the switching period:

$$i_L(t) = \frac{V_e}{L}t \quad [1.17]$$

giving a maximum current $I_{L\max}$ at the end of the first phase, which is expressed as:

$$I_{L\max} = \frac{\alpha.T_d.V_e}{L} \quad [1.18]$$

In the second operating phase, over the time interval $[\alpha T_d, T_d]$, the transistor is off. Consequently:

$$i_T = 0 \quad [1.19]$$

and as the current i_L in the inductance is non-null (and has a value of I_{Lmax} at the beginning of this phase), the diode must enter into conduction to ensure current continuity:

$$i_d = i_L \quad [1.20]$$

hence:

$$V_d = 0 \quad [1.21]$$

This means that the voltage at the inductance terminals is equal to $-V_s$:

$$V_L = -V_s \quad [1.22]$$

hence:

$$i_L(t - \alpha T_d) = I_{L\max} - \frac{V_s}{L}(t - \alpha T_d) \qquad [1.23]$$

In critical conduction, the continuous mode hypotheses are still applicable. Thus:

$$\langle V_L \rangle = 0 \qquad [1.24]$$

As the instantaneous expression of V_L is known for the whole switching period:

$$\langle V_L \rangle = \frac{1}{T_d}(\alpha T_d.V_e - (1-\alpha).T_d.V_s) \qquad [1.25]$$

and finally:

$$V_s = \frac{\alpha.V_e}{1-\alpha} \qquad [1.26]$$

Using this expression, we have:

– $V_s = V_e$ for $\alpha = 1/2$;

– $V_s < V_e$ for $\alpha < 1/2$;

– $V_s > V_e$ for $\alpha > 1/2$.

This conforms precisely to our initial predictions.

Let us now consider the average value of the current absorbed from the input voltage source, noting that the current absorbed by the converter is equal to the current in the transistor. It is therefore:

– equal to $\frac{\alpha T_d V_e}{L}$ during the first phase of the switching period;

– equal to 0 during the second phase.

Finally, note that there is no third phase if the converter is kept in a state of critical conduction. Consequently, the average value $\langle i_L \rangle$ of i_L may be calculated as follows:

$$\langle i_L \rangle = \frac{\alpha^2 T_d V_e}{2L} \quad [1.27]$$

Note that at low-frequency levels (in terms of the average over the switching period), the current absorbed by the converter is proportional to the voltage V_e. An equivalent resistance R_{eq} can therefore be introduced of the form:

$$R_{eq} = \frac{2L}{\alpha^2 T_d} \quad [1.28]$$

Thus, if the chopper is powered using a diode rectifier, it will enable absorption of a sinusoidal current in the network, in the same way as the boost converter presented in the previous section. In this case, this behavior is obtained naturally as long as the converter remains in a state of critical conduction. Power factor correctors (PFCs) of this type are therefore much simpler to control than the boost converters presented in Chapter 3 of Volume 2 [PAT 15b].

Critical operation is at the limits of continuous conduction, and so the converter input/output relationship in this case is that established in [1.26], for any load (on the condition that I_s is sufficiently high to ensure continuous conduction). However, we now need to consider the case of discontinuous conduction. The basic principles of this study were covered in the context of machine power supplies: the average converter input and output powers over a switching period need to be identified, and these values must be equal to enable operation in permanent mode (even in the cases where the converter includes an internal storage element, as in this case). Thus, the output voltage P_s may simply be written as follows:

$$P_s = V_s.I_s \quad [1.29]$$

where V_s and I_s are presumed to be strictly constant (leaving aside all ripple, notably in the voltage). We now need to evaluate the average converter input power P_e, which is the product of V_e (presumed to be strictly constant) and the average value of the input current I_e (denoted as i_T in the diagram shown in Figure 1.5). This current may easily be expressed as a function of time, as we know that:

$$i_T = i_L \, \forall t \in [0; \alpha.T_d] \quad [1.30]$$

and:

$$i_T = 0 \, \forall t \notin [0; \alpha.T_d] \quad [1.31]$$

As current i_L is null at $t = 0$ (discontinuous mode operation), we may write:

$$i_T = i_L = \frac{V_e}{L} t \, \forall t \in [0; \alpha.T_d] \quad [1.32]$$

hence:

$$I_e = \langle i_T \rangle = \frac{\alpha^2.V_e}{2L.F_d} \quad [1.33]$$

and thus:

$$V_s.I_s = \frac{\alpha^2.V_e^2}{2L.F_d} \quad [1.34]$$

hence:

$$V_s = \frac{\alpha^2.V_e^2}{2L.F_d.I_s} \quad [1.35]$$

Note that the converter is a power source controlled by the duty ratio α. A set of hyperboles is thus obtained for the output voltage V_s as a function of the charge current I_s.

As before, it is best to work with reduced quantities, denoted as $y = V_s/V_e$ and $x = 2L.F_d.I_s/E$. Hence:

$$y = \frac{\alpha^2}{x} \qquad [1.36]$$

To establish the equation of the critical location of this converter, α must be eliminated from the equation, using the relationship $y(x)$ obtained in continuous (or critical) conduction, insofar as the critical location is the geometric zone which simultaneously corresponds to both operating modes of the converter. We therefore rewrite relationship [1.26] using reduced variables:

$$y = \frac{\alpha}{1-\alpha} \qquad [1.37]$$

Inverting this equation gives:

$$\alpha = \frac{y}{1+y} \qquad [1.38]$$

Replacing α with this expression in [1.36], then:

$$x = \frac{y}{1+y^2} \qquad [1.39]$$

This may be approximated to a linear function for $y^2 \ll 1$:

$$x \underset{0}{\sim} y \qquad [1.40]$$

It tends asymptotically toward a hyperbole when $y^2 \gg 1$:

$$x \underset{\infty}{\sim} \frac{1}{y} \qquad [1.41]$$

This characteristic is well known in electrical engineering, and is analogous to the couple versus sliding coefficient relationship of an induction machine. The obtained curve (see Figure 1.6) is known as a serpentine. Note the maximum value of x, denoted as $x_{max} = 1/2$, in this critical location (corresponding to $y = y_{max} = 1$).

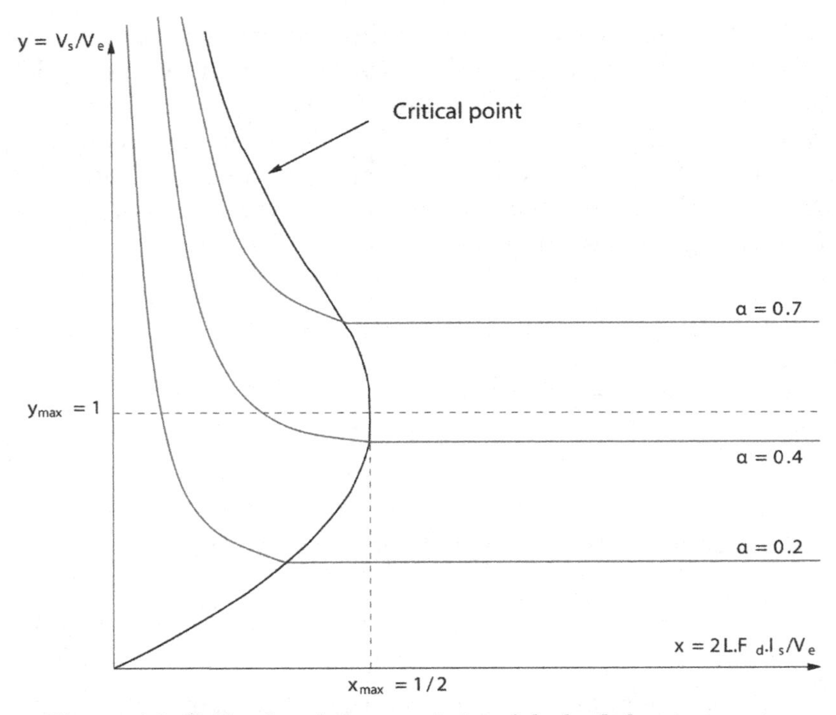

Figure 1.6. *Full reduced characteristics of the buck–boost converter (continuous and discontinuous conduction)*

The waveforms corresponding to discontinuous conduction in the buck–boost converter are shown in Figure 1.7 for illustrative purposes. Note that the only difference between these waveforms and those obtained in continuous conduction is the absence of a third phase in the switching period (this is punctual in the borderline case of critical conduction).

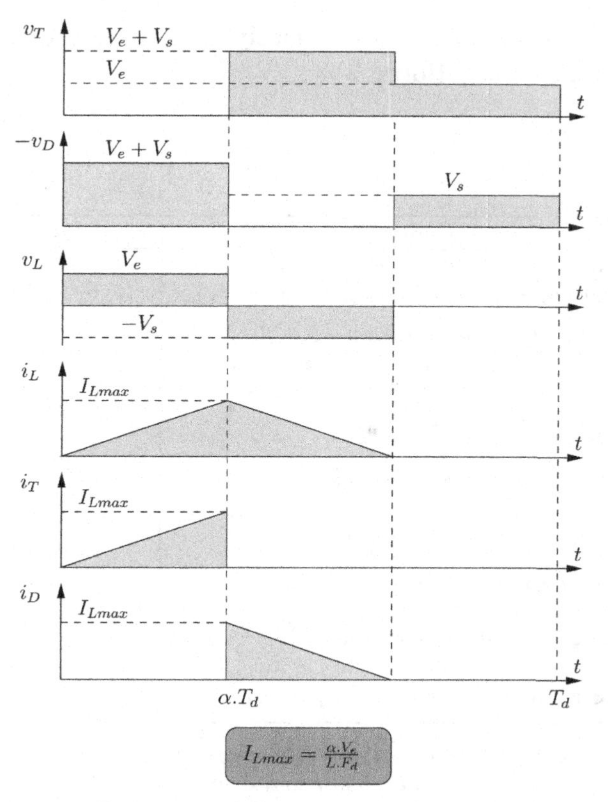

Figure 1.7. *Waveforms of electrical quantities in the buck–boost converter (discontinuous conduction)*

To complete our study of the buck–boost converter, it is useful to establish a summary of the quantities involved in the dimensioning process (see Table 1.3). This summary corresponds to the case of critical conduction, and in this case, we still need to calculate the output voltage ripple in order to calculate the capacitance C required to fulfill a given specification. As in the case of the boost converter, the capacitor in this converter discharges with a constant current when the transistor is in a state of conduction. We therefore observe the same behavior with regard to the voltage ripple (note that this formula is not, strictly speaking, applicable for discontinuous conduction):

$$\Delta V_s = \frac{\alpha.I_s}{C.F_d} \qquad [1.42]$$

Quantities	*Values*
Maximum transistor voltage V_{Tmax}	$V_e + V_s$
Maximum inverse diode voltage V_{dmax}	$-(V_e + V_s)$
Current ripple in inductance Δi_L	$\frac{\alpha.V_e}{L.F_d}$
Average output voltage $\langle V_s \rangle$	$\frac{\alpha.V_e}{1-\alpha}$
Output voltage ripple Δv_s	$\frac{\alpha.I_s}{C.F_d}$
Maximum current in transistor and diode	$\frac{\alpha.V_e}{L.F_d}$
Average current in transistor $\langle I_T \rangle$	$\frac{\alpha^2.V_e}{2L.F_d}$
Average current in diode $\langle I_d \rangle$	$\frac{\alpha(1-\alpha).V_e}{2L.F_d}$
RMS current in transistor (for $\Delta i_L \ll \langle i_L \rangle$)	$\frac{\alpha.V_e}{L.F_d} \cdot \sqrt{\frac{\alpha}{3}}$
RMS current in diode (for $\Delta i_L \ll \langle i_L \rangle$)	$\frac{\alpha.V_e}{L.F_d} \cdot \sqrt{\frac{1-\alpha}{3}}$

Table 1.3. *Summary of critical conduction in the buck–boost converter*

2

Isolated Converters

2.1. Forward converters

A forward converter is an isolated version of the buck converter in which voltage bursts are sent to an isolated transformer, placed between the "switching cell" and the LC output filter. However, there is, *a priori*, a problem concerning magnetization of the transformer due to the application of a non-null average voltage to the primary winding. A demagnetization circuit is needed to counteract this effect: several variations may be used (dissipative or non-dissipative), but in this case, we will only consider the structure shown in Figure 2.1, using a third winding for demagnetization purposes.

Therefore, the transformer can be modeled using three ideal windings (resistances are neglected) and with a magnetizing inductance localized[1] at the primary winding terminals. The magnetic coupling between the three windings is also supposedly ideal (no leakage inductance).

As for previous converters, we decompose the switching period T_d in relation to the control of transistor T, presumed

1 Further details of transformer modeling are given in Chapter 5 of Volume 1 [PAT 15a].

to be in a state of conduction over the interval $[0, \alpha.T_d]$. For this phase, we may write:

$$v_1 = V_e \quad [2.1]$$

hence:

$$v_2 = m.V_e \quad [2.2]$$

with $m = n_2/n_1$. In the same way, we have:

$$v_3 = m'.V_e \quad [2.3]$$

with $m' = n_3/n_1$. Note that:

$$v_{D3} = -\left(1 + m'\right).V_e < 0 \quad [2.4]$$

This shows that diode D_3 is reverse biased, and thus switched off.

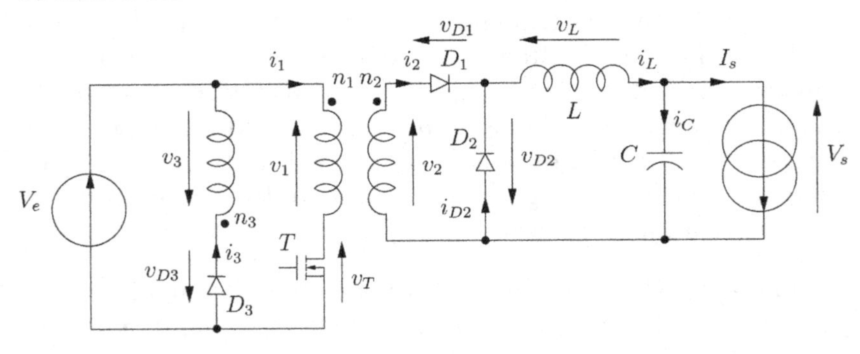

Figure 2.1. *Diagram of a "forward" power supply*

Considering the secondary (principal) of the transformer, we see that the association of D_1 and D_2 constitutes a rectifier, allowing passage of the highest positive tension between zero and v_2. As we have seen, in this phase, $v_2 = m.V_e > 0$; diode D_1 is consequently switched on ($v_{D1} = 0$), and diode D_2 will be switched off ($v_{D2} = -m.V_e < 0$). In this case, we will presume that the rectifier (D_1, D_2) is operating in continuous conduction mode, and that the current i_L does not cancel out during the switching period T_d. In these conditions, we have:

$$v_L = m.V_e - V_s \quad [2.5]$$

Note that in the case where diode D_2 is on, the voltage v_L will be equal to $-V_s$, and therefore clearly negative. Consequently, given that permanent periodical mode imposes an average voltage of zero at the inductance terminals, we deduce that $m.V_e - V_s$ is positive.

Based on the usual hypothesis that $V_s = C^{te}$, an expression of $i_L(t)$ may be established, presuming that the current has a total value denoted as i_{Lmin} at the beginning of the phase (this is a minimum value, as, due to the sign of v_L, we have established that the current is increasing):

$$i_L(t) = \frac{m.V_e - V_s}{L} t + i_{Lmin} \quad [2.6]$$

In the same way, the magnetizing current $i_{mag}(t)$ in the transformer primary can be expressed by noting that:

$$i_1(t) = m.i_2(t) + i_{mag}(t) \quad [2.7]$$

with:

$$i_{mag}(t) = \frac{V_e}{L_{mag}} t \quad [2.8]$$

This includes a magnetizing inductance L_{mag} which still needs to be defined (see Chapter 5 of Volume 1 [PAT 15a]). Moreover, note that the magnetizing current is null at $t = 0$, as a correctly dimensioned transformer and converter will guarantee full demagnetization at the end of the period T_d.

Transistor ($i_1 = 0$) is switched off at instant $t = \alpha.T_d$, and the continuity of the magnetic flux in the transformer (and of the magnetizing current i_{mag}) means that diode D_3 enters a state of conduction. This gives us:

$$v_3 = -V_e \qquad [2.9]$$

$$v_1 = -\frac{V_e}{m'} \qquad [2.10]$$

$$v_2 = -\frac{n_2}{n_3}V_e = -\frac{n_2}{n_1} \cdot \frac{n_1}{n_3}V_e = -\frac{m}{m'}V_e \qquad [2.11]$$

It is easy to verify that diode D_1 is off by the inversed sign of voltage v_2. Continuity in current i_L means that diode D_2 must enter into conduction, hence:

$$v_L = -V_s \qquad [2.12]$$

We still need to determine the conduction period for diode D_3, used to fully demagnetize the transformer before the end of the switching period. To do this, we note that the magnetizing inductance L_{mag} of the transformer has been localized on the primary winding (n_1); we can then calculate the evolution of i_{mag} from its maximum value (reached at the end of the previous phase, at $t = \alpha.T_d$), denoted as i_{mag}^{max}:

$$i_{mag}(t) = i_{mag}^{max} - \frac{V_e}{m'.L_{mag}}t \qquad [2.13]$$

REMARK 2.1.– Note that the time origin point has been changed to simplify the notation. Instant $t = 0$ in the expression of $i_{mag}(t)$ presented above corresponds to the final instant of the previous phase (i.e. $t = \alpha.T_d$ in relation to the initial time origin).

Based on this result, we may write that, for full demagnetization to occur ($i_{mag} = 0$), we must wait for instant:

$$t = \frac{m'.L_{mag}.i_{mag}^{max}}{V_e} \quad [2.14]$$

As we know that i_{mag}^{max} is reached at the end of the first phase (expression [2.8] of i_{mag} for $t = \alpha.T_d$), we obtain:

$$t = m'.\alpha.T_d \quad [2.15]$$

We know that this instant must occur at the end of the switching period, at the latest. Thus, we obtain the following inequality:

$$\alpha.\left(1 + m'\right) \leq 1 \quad [2.16]$$

The control signal must therefore have a maximum duty ratio α_{max} which is a function of the turns ration $m' = n_3/n_1$:

$$\alpha_{max} = \frac{1}{1 + m'} \quad [2.17]$$

For example, taking $n_1 = n_3$ (a very widespread situation, simple to carry out for a "forward" transformer), the maximum duty ratio will be equal to $1/2$.

Based on this operation, we can introduce a final phase during which:

– transistor T remains off;

– diode D_3 switches off once the transformer is fully demagnetized;

– diode D_1 remains off;

– diode D_2 remains on (the continuous conduction hypothesis remains valid).

In these conditions, note that:

$$v_1 = v_2 = v_3 = 0 \qquad [2.18]$$

and, clearly:

$$i_1 = i_2 = i_3 = 0 \qquad [2.19]$$

Finally, the rectifier (D_1, D_2) and the LC output filter operate in exactly the same way as for the second phase:

$$v_L = -V_s \qquad [2.20]$$

We can, therefore, use the relationship:

$$\langle v_L \rangle = 0 \qquad [2.21]$$

for the whole of the switching period in order to determine the converter input/output relationship:

$$V_s = m.\alpha.V_e \qquad [2.22]$$

2.2. Flyback converters

The flyback converter, as shown in Figure 2.2, is an adaptation of the buck–boost converter where the storage winding is split into two windings in order to obtain galvanic isolation. Note that the transformer used in this context operates in a way which is fundamentally different from that used in the forward converter: an air gap is used in order to store energy, leading to the production of a strong magnetizing current. There is no direct energy transfer between the primary (n_1) and the secondary (n_2); instead, we observe two distinct phases:

– a storage phase, where energy is retained in the magnetic circuit by the primary winding (for the interval $[0; \alpha.T_d]$);

– a release phase, where this energy is released by the secondary winding (for the remainder of the period).

The interest of operating in critical mode is evident, even before calculations (full energy release at the exact end of the switching period T_d), as for the buck–boost converter; this enables optimal use of both the "transformer" and the switches. In this context, we will not consider the continuous and discontinuous conduction modes of this conductor, as these points were subject to exhaustive coverage for the buck–boost converter (these results are similar for non-isolated converters and their isolated variations).

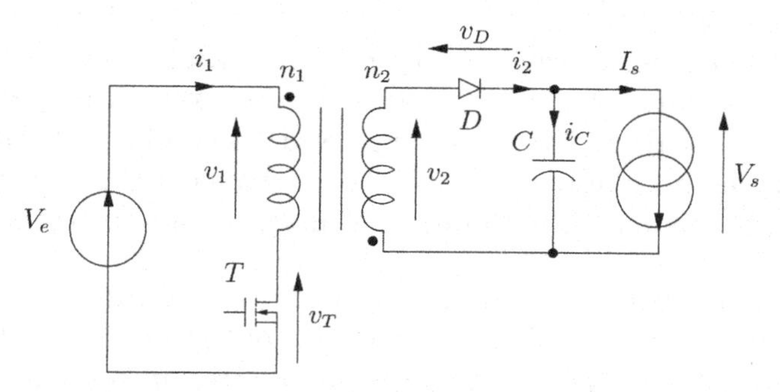

Figure 2.2. *Diagram of a flyback converter*

Based on the hypothesis of full demagnetization, we may consider that at $t = 0$, as soon as the transistor enters a state of conduction, the current i_1 is null and will increase as a function of time, in proportion to the voltage applied to the primary winding. In this case, we have:

$$v_1 = V_e \quad [2.23]$$

Considering that the winding presents an eigen-inductance L_1, we have:

$$i_1(t) = \frac{V_e}{L}t \quad [2.24]$$

The flux in the magnetizing circuit is a function of this current alone, as, due to the orientation of the "secondary" winding (see the notion of "corresponding points" for magnetically coupled windings in Chapter 5 of Volume 1 [PAT 15a]), we have a negative voltage v_2 (taking $m = n_2/n_1$):

$$v_2 = -m.V_e \qquad [2.25]$$

Consequently, the voltage v_D applied to the diode terminals is negative:

$$v_D = -m.V_e - V_s < 0 \qquad [2.26]$$

The diode is therefore in the off state, leading us to stipulate that $i_2 = 0$. There are no simultaneous contributions from the "primary" and "secondary" windings in terms of the magnetic flux in the circuit during this phase (this is also true for the following phase, as we will see).

When the transistor is in the off state, at the end of the first phase (at $t = \alpha.T_d$), the continuity of the magnetic flux in the "transformer" requires diode D to enter into conduction. Current discontinuity is possible (at least in the context of primary analysis) in windings placed on a magnetic circuit shared with other windings, as it is the magnetic energy which cannot tolerate discontinuity, and this energy is written, in this context:

$$W_{mag} = \frac{1}{2}\sum_{p,q} \psi_p i_q \qquad [2.27]$$

Thus, at the moment of switching, the fact that $L_1.i_1$ drops to zero is offset by the appearance of $L_2.i_2$, which was null during the first phase. The magnetic energy before switching may be expressed as:

$$W_{mag}\left(\alpha.T_d^-\right)=\frac{1}{2}L_1.i_1\left(\alpha.T_d^-\right)^2=\frac{1}{2}L_1\left(\frac{V_e.\alpha.T_d}{L_1}\right)^2$$
$$=\frac{1}{2L_1}\left(V_e.\alpha.T_d\right)^2 \quad [2.28]$$

This energy is transferred during the second phase of the switching period (between $\alpha.T_d$ and T_d). Generally speaking, the behavior of this converter is very simple to analyze from an energy perspective; at $t=\alpha.T_d^+$, we therefore have:

$$i_2\left(\alpha.T_d^+\right)^2=\frac{L_1}{L_2}i_1\left(\alpha.T_d^-\right)^2=\frac{L_1}{L_2}\left(\frac{V_e.\alpha.T_d}{L_1}\right)^2=\frac{\alpha^2V_e^2}{L_1L_2F_d^2} \quad [2.29]$$

It is important to note that the relationship between inductances L_1 and L_2 is equal to the square of the relationship between the number of turns n_1 and n_2 (this result is demonstrated in Chapter 5 of Volume 1 [PAT 15a]). By recalling the turns ratio $m = n_2/n_1$ shown above, we obtain:

$$i_2\left(\alpha.T_d^+\right)=\frac{i_1\left(\alpha.T_d^-\right)}{m} \quad [2.30]$$

The notion of energy conservation can, in fact, be replaced in the context of magnetically coupled windings using the notion of conservation of "ampere-tours", as the equality $\frac{1}{2}L_1.i_1^2=\frac{1}{2}L_2.i_2^2$ can be replaced by $n_1.i_1=n_2.i_2$.

In conclusion to this initial "magnetic" study of the flyback converter, note that the level of modeling used for the "transformer" in this context is sufficient for operational analysis; however, it needs to be refined for design purposes, taking account of a parasitic phenomenon in the form of the imperfect connection between the two windings and the existence of magnetic leaks. This phenomenon means that the structure presented in Figure 2.2 cannot be used in practice without the addition of an auxiliary element, known as a snubber, which will be covered in section 2.5.

First, however, it is useful to return to the characterization of the output voltage of the converter; operating in critical mode, the input/output relationship should be analogous to that of the buck–boost converter (see equation [1.26]) from which it derives. To do this, we note that the average voltage at the terminals of the winding (primary or secondary) is zero for the switching period. We have presumed that current i_2 only cancels out for brief periods at the end of the switching period (by definition in critical mode): diode D, therefore, continues to conduct for the whole interval $[\alpha.T_d; T_d]$. In these conditions, voltage V_s is applied to the "secondary" winding for the whole interval; as we demonstrated in equation [2.25], this voltage is equal to $m.V_e$ during the interval $[0; \alpha.T_d]$. We can then calculate the expression of the average value of v_2:

$$\langle v_2 \rangle = \frac{1}{T_d} \left(-\alpha.T_d.m.V_e + (1-\alpha).T_d.V_s \right) \qquad [2.31]$$

Noting that this average value should be null in permanent mode, the following relationship can be deduced:

$$V_s = \frac{m.\alpha.V_e}{1-\alpha} \qquad [2.32]$$

This clearly corresponds to the relationship obtained for a buck–boost converter, up to the turns ratio m.

2.3. Dimensioning a flyback transformer

In dimensioning a flyback transformer, we begin by noting that the stored magnetic energy is fundamental to the operation of the converter. This element is not a true transformer, but rather inductances connected by a shared magnetic circuit including an air gap. When dimensioning the "transformer", we therefore need to consider:

– a ferrite core reference;

– the dimensions of the primary and secondary windings;

– the air gap.

These choices are made based on an analysis of the operation of the converter, and on the impact of this operation on the magnetic states of the ferrite core and the windings. In this case, the hypothesis of full demagnetization is used; as discussed in Chapter 1, this is the best operating mode for this type of converter (as for buck–boost converters). The maximum magnetic flux in the core (of section A_e) can, therefore, be expressed as:

$$n_1.B_{max}.A_e = V_e.\alpha.T_d \qquad [2.33]$$

hence:

$$A_e = \frac{\alpha.V_e}{n_1.B_{max}.F_d} \qquad [2.34]$$

The dimensions of the windings are then determined in terms of the conductor cross-section. To do this, note that (for $0 \leq t \leq \alpha.T_d$):

$$i_1(t) = \frac{V_e}{L_1} \cdot t \qquad [2.35]$$

From this, the root mean square (RMS) value $I_{1\text{RMS}}$ of the current can be deduced:

$$I_{1\text{RMS}} = \sqrt{\frac{1}{T_d} \cdot \int_0^{\alpha.T_d} \left(\frac{V_e}{L_1} \cdot t\right)^2 dt} = \sqrt{\frac{1}{T_d} \cdot \left(\frac{V_e}{L_1}\right)^2 \cdot \left[\frac{t^3}{3}\right]_0^{\alpha.T_d}}$$

$$= \frac{\alpha\sqrt{\alpha}V_e}{\sqrt{3}L_1.F_d} \qquad [2.36]$$

The secondary of this transformer is traversed by a current with an amplitude I_{2max} linked to that of the primary current, denoted as I_{1max}, by the relationship:

$$n_1.I_{1max} = n_2.I_{2max} \qquad [2.37]$$

hence:

$$I_{2max} = \frac{n_1}{n_2} I_{1max} = \frac{I_{1max}}{m} \qquad [2.38]$$

When dimensioning in the context of critical conduction with a maximum charge current, in this case, the characteristic point of the buck–boost converter appears ($x = 1/2$; $y = 1$), for which $\alpha = 1/2$. In these conditions, the expression of $i_2(t)$ in critical conduction corresponds to a symmetrical waveform of $i_1(t)$ in relation to the mid-point of the switching period (up to the turns ratio m); thus, in this case:

$$I_{1\text{RMS}} = \frac{V_e}{2\sqrt{6}L_1.F_d} \qquad [2.39]$$

and:

$$I_{2\text{RMS}} = \frac{V_e}{2\sqrt{6}m.L_1.F_d} \qquad [2.40]$$

Hence:

$$S_b = \frac{n_1.I_{1RMS} + n_2.I_{2RMS}}{K_b.J_{max}} = \frac{(n_1 + n_2/m).V_e}{2\sqrt{6}L_1.F_d.K_b.J_{max}}$$
$$= \frac{n_1.V_e}{\sqrt{6}L_1.F_d.K_b.J_{max}} \qquad [2.41]$$

The product $A_e.S_b$ of the selected ferrite core must satisfy the following inequality (α has been replaced by $1/2$ in the expression of A_e):

$$A_e.S_b \geq \frac{V_e^2}{2\sqrt{6}B_{max}.K_b.J_{max}.L_1.F_d^2} \qquad [2.42]$$

Moreover, this expression can be modified to include the notion of power $\mathcal{P}$ drawn from the input source insofar as:

$$\mathcal{P} = V_e.\langle i_1 \rangle = \frac{\alpha^2.V_e^2.T_d}{2L_1} \qquad [2.43]$$

hence, for $\alpha = 1/2$:

$$\mathcal{P} = V_e.\langle i_1 \rangle = \frac{V_e^2.T_d}{8L_1} = \frac{V_e^2}{8L_1.F_d} \qquad [2.44]$$

thus:

$$A_e.S_b \geq 2\sqrt{\frac{2}{3}} \cdot \frac{\mathcal{P}}{B_{max}.K_b.J_{max}.F_d} \qquad [2.45]$$

The choice of ferrite core is, therefore, not directly linked to the windings, but simply to the power needing to be transferred (and thus, for a given switching frequency, to the energy stored in the air gap). Once a core has been chosen, a standardized air gap is selected, giving a certain specific inductance a_L, for which:

$$L_1 = a_L.n_1^2 \qquad [2.46]$$

Inductance L_1 is defined based on the expression of the power [2.43], allowing the required number of turns n_1 to be determined; the turns ratio needed for the specific application allows us to determine the number of turns n_2.

Unlike smoothing inductances (as used at the output point of a buck converter or a forward power supply), the currents circulating in the primary and secondary windings of a flyback transformer display a high level of ripple. In these conditions, the skin thickness should be greater than or equal to the radius of a conductor. To do this, the following inequality must be verified:

$$\sqrt{\frac{2}{\omega\mu\sigma}} \geq \frac{I_{k\mathrm{RMS}}}{J_{max}} \qquad [2.47]$$

for $k = 1$ or 2 (primary or secondary). If this condition is not satisfied, the use of litz wire (see Figure 2.3) is highly recommended. This type of conductor is made up of a large number of fine strands, each isolated from the others, and is used for high frequency (HF) currents.

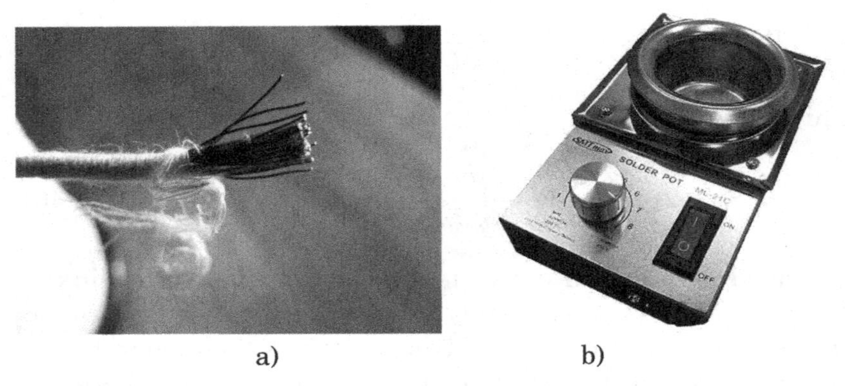

a) b)

Figure 2.3. *Litz wire a) and tin bath for tinning extremities b)*

Note, however, that specific equipment is required in order to use this type of wire correctly. As each strand is isolated from the others, each strand must be electrically connected to enable connections at the extremities of the winding. This is generally achieved by soaking the wire in a bath of molten tin in order to eliminate the insulating varnish and solder the strands together correctly. If this process is not carried out, only part of the cross-section of the conductor may effectively be used, and the maximum acceptable current density may be exceeded in those strands which are correctly connected.

REMARK 2.2.– The storage winding of a buck–boost converter should be dimensioned using the same approach as for coupled inductances. The operation of this winding should not be compared with that of the smoothing inductances used for the output of buck and forward converters. In this context, even the conductor technology needs to be evaluated, as although the current is continuous in a single winding of this type, it is subject to high ripple levels, and the alternating component is no longer negligible in relation to the continuous component.

2.4. Dimensioning a forward transformer

An important relationship relating to the operation of the converter, defined in section 2.1, is essential when dimensioning a forward transformer:

$$\alpha_{max} = \frac{1}{1+m'} \qquad [2.48]$$

with $m' = n_3/n_1$.

The primary (n_1) and demagnetization (n_3) windings are conventionally created together in order to ensure the best possible coupling: thus, we have $n_1 = n_3$. In these conditions, we obtain a maximum duty ratio of $\alpha_{max} = 1/2$. Moreover, note that the maximum magnetic flux created in the primary winding is expressed as:

$$n_1 \cdot B_{max} \cdot A_e = V_e \cdot \frac{T_d}{2} \qquad [2.49]$$

hence:

$$A_e = \frac{V_e}{2.n_1.B_{max}.F_d} \qquad [2.50]$$

We then dimension the conductors, noting that the dimensioning quantity is the RMS value of the current. If we consider that the current in the smoothing inductance placed at the output point of the rectifier on the secondary of the transformer (n_2) is quasi-constant, the waveforms of the primary and secondary currents i_1 and i_2 may be seen as bursts, varying between 0 and the value $m.I_s$ (for the primary, with $m = n_2/n_1$) and I_s (for the secondary) with non-null plateaux of width $\alpha.T_d$ (giving a maximum width of $T_d/2$). Consequently, the primary and secondary RMS currents I_{1RMS} and I_{2RMS} are expressed as:

$$\begin{cases} I_{1\text{RMS}} = m.I_s/\sqrt{2} \\ I_{2\text{RMS}} = I_s/\sqrt{2} \end{cases} \quad [2.51]$$

Given that the current density J_{max} and a winding coefficient K_b are specified, we are then able to deduce the required winding cross-section. However, the dimensioning process must take account of the winding of the magnetization winding. This winding is only traversed by the magnetization current during intervals with a duration shorter than or equal to $(1-\alpha).T_d$. As this current is weak, the cross-section of wire required for this function is negligible. In practice, however, wire with the same cross-section as for the primary winding is used in order to guarantee strong coupling between the windings (the number of turns in each winding is also identical for the same reason). In these conditions, we obtain the following result:

$$S_b = \frac{2n_1.m.I_s + n_2.I_s}{K_b.J_{max}.\sqrt{2}} = \frac{3n_1.m.I_s}{K_b.J_{max}.\sqrt{2}} \quad [2.52]$$

Finally, a ferrite core reference suitable for the specification must be selected, i.e. with a product $A_e.S_b$ greater than the product of the expressions determined above for each of the relevant factors:

$$A_e.S_b \geq \frac{3.m.V_e.I_s}{2\sqrt{2}.B_{max}.F_d.K_b.J_{max}} \quad [2.53]$$

First, note that the number of primary turns n_1 is no longer explicitly included in this expression (except in the transformation ratio $m = n_2/n_1$). Moreover, note that power $\mathcal{P}$ drawn from the source may be introduced into this expression, as:

$$\mathcal{P} = V_e. \langle i_1 \rangle = \alpha.mV_eI_s \qquad [2.54]$$

As surface A_e has already been determined for the most critical case ($\alpha_{max} = 1/2$), this can be rewritten as:

$$\mathcal{P} = \frac{m.V_e.I_s}{2} \qquad [2.55]$$

hence:

$$A_e.S_b \geq \frac{3.\mathcal{P}}{\sqrt{2}.B_{max}.F_d.K_b.J_{max}} \qquad [2.56]$$

While product $A_e.S_b$ is neither a surface nor a volume, it gives us an indication of the physical size of the transformer. From the expression shown above, we see that this "size" is inversely proportional to the operating frequency F_d.

Once a core has been selected, cross-sections A_e and S_b are defined, and, using A_e and equation [2.50], it is possible to determine the number of primary turns n_1. Once this parameter has been determined, we know that the number of turns n_3 is identical, and the turns ratio imposed by the application may be used to deduce n_2 directly from n_1.

2.5. Snubbers

Snubbers (as used in switching assistance circuits) are auxiliary circuits used in isolated converters to protect controlled switches from overvoltages, which may result from imperfect coupling of windings (i.e. leakage inductance). To demonstrate the necessity of these circuits, we must return to the modeling of coupled inductances discussed in Chapter 5 of Volume 1 [PAT 15a]. It is entirely natural to use snubber circuits in forward and flyback power supplies. We will describe our dimensioning methodology based on this second structure.

2.5.1. *Impact of transformer leakage inductance in a converter*

Chapter 5 of Volume 1 [PAT 15a] included an equivalent diagram of coupled inductances, showing three parameters:

– the magnetizing inductance L_1 localized in the primary winding;

– the leakage inductance $\sigma.L_2$ totaled in the secondary winding;

– the turns ratio $m = M/L_1$ between the primary and secondary windings.

This is not the only possible modeling; it is also possible to work with the initial equations of the coupled inductances, working from the secondary to the primary, displaying the magnetizing inductance in the secondary and totaling losses in the primary. We will take this approach here. First, we note the following equations:

$$v_1 = n_1 \frac{d\Psi_1}{dt} = L_1 \frac{di_1}{dt} + M \frac{di_2}{dt} \qquad [2.57]$$

and:

$$v_2 = n_2 \frac{d\Psi_2}{dt} = M \frac{di_1}{dt} + L_2 \frac{di_2}{dt} \qquad [2.58]$$

Equation [2.58] is then factorized by L_2 to obtain:

$$v_2 = L_2 \left(\frac{di_2}{dt} + \frac{M}{L_2} \cdot \frac{di_1}{dt} \right) \qquad [2.59]$$

This gives a new turns ratio $m' = M/L_1$; ideally, this ratio should also be equal to the "inverse" ratio of the number of turns in the two windings ($n_1/n_2 = 1/m$). Hence:

$$v_2 = L_2 \frac{d}{dt}\left(i_2 + m'.i_1\right) \qquad [2.60]$$

As in Chapter 5, Volume 1 [PAT 15a], the next step is to calculate the expression $m'.v_2$ (with a simple inversion of roles between the primary and the secondary):

$$m'.v_2 = \frac{M}{L_2}\left(M\frac{di_1}{dt} + L_2\frac{di_2}{dt}\right) = M\frac{di_2}{dt} + \frac{M^2}{L_2}\cdot\frac{di_1}{dt} \qquad [2.61]$$

This equation includes a term $M\frac{di_2}{dt}$, which has already been encountered in the expression of v_1. This latter expression may be rewritten as follows:

$$v_1 = m'.v_2 - \frac{M^2}{L_2}\cdot\frac{di_1}{dt} + L_1\frac{di_1}{dt} = m.v_1 + \left(1 - \frac{M^2}{L_1 L_2}\right)L_1\frac{di_1}{dt} \qquad [2.62]$$

This gives the dispersion coefficient σ, representative of losses, and this time the leakage inductance $\sigma.L_1$ is totaled in the primary. The corresponding equivalent diagram is shown in Figure 2.4.

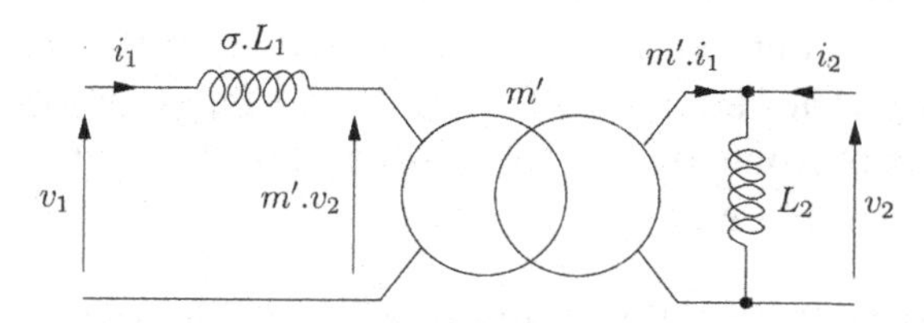

Figure 2.4. *Equivalent electrical diagram of a transformer with losses totaled in the primary*

2.5.2. *Implementation and dimensioning of a snubber*

When the input transistor of a flyback converter is opened, the primary current cancels out immediately in the case of an idea coupling since, as we have already seen, the important state variable of this type of coupled winding system is the magnetic energy stored in the air gap, and not the primary and secondary current. Unfortunately, if we look more closely, magnetic leakage means that the primary current is still a state variable which will not tolerate discontinuity. As the primary current is subject to a rapid drop imposed by the transistor receiving the order to switch off, an overvoltage is generated by the leakage inductance, denoted as $\sigma.L_1$ in the diagram shown in Figure 2.2. This is simply denoted by l_f in the diagram of the flyback converter in Figure 2.5.

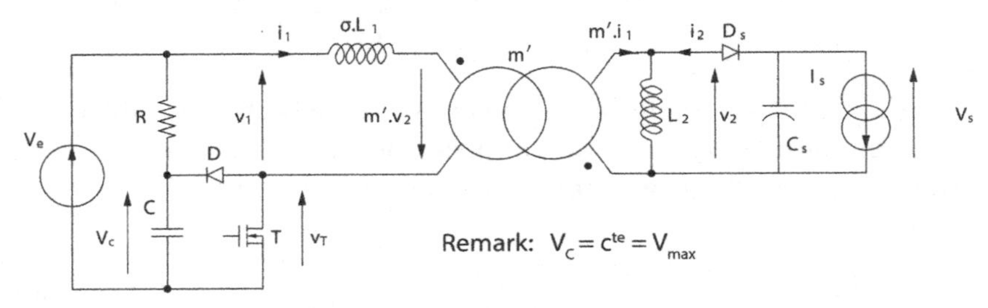

Figure 2.5. *Flyback converter with RCD snubber*

This overvoltage is therefore driven only by the switching speed of the transistor. This situation is not generally acceptable (except when using the avalanche phenomenon present in metal oxide semiconductor field effect transistor (MOSFET), which can potentially withstand this type of stress[2]) and the energy $W_f = 1/2.l_f.I_{1max}^2$ stored in the leakage inductance needs to be dissipated in an additional circuit: the snubber.

This function can be provided in different ways:

– inclusion of a *transil* diode in parallel to the transistor (chosen to dissipate the required power $W_f.F_d$ and to prevent the voltage at the transistor terminals from exceeding a certain voltage $V_{max} > V_e + V_s/m$);

2 This subject is covered in manufacturer application notes.

– the use of a resistance + capacitor + diode (RCD)-type snubber.

This second structure is shown in Figure 2.5. Note that the snubber circuit associated with transistor T in the converter and with the leakage inductance l_f has a structure similar to the boost converter. The voltage obtained at the capacitor terminals is higher than the input voltage, with a value of $V_e + V_s/m$. We, therefore, begin by choosing the maximum voltage to be imposed on the transistor (generally half of its caliber, for example 300 V for a 600 V transistor) and ensure that the resistance R will dissipate the power $W_f.F_d$ at this voltage:

$$R = \frac{(V_{max} - V_e)^2}{W_f.F_d} \tag{2.63}$$

We then select a smoothing capacitor C to smooth the voltage; the time constant $R.C$ must, therefore, be high in relation to the switching period $T_d = 1/F_d$ used for the converter.

In both cases (transil or RCD circuit), good design practice is based on the knowledge of the leakage inductance l_f. This information must be obtained by experimental testing, generally using traditional transformer testing techniques (no-load tests, and more specifically short-circuit tests, in this case for the leakage inductance). To do this, an arbitrary waveform generator (AWG) may be tuned as a sinusoidal voltage source at a frequency close to that which will eventually be used for switching.

3

Resonant Converters and Soft Switching

3.1. Soft switching

3.1.1. *Definitions, ZVS and ZCS switching*

Soft switching consists of eliminating all switching losses by cancellation of the instantaneous power devices at the exact moment of turn-on or turn-off. The losses p_K in a switch K take the form:

$$p_K = v_k . i_K \qquad [3.1]$$

where v_K and i_K are the voltage and the current at the component terminals. In order to eliminate losses, we must either cancel the voltage (*zero voltage switching* (ZVS)) or the current (*zero current switching* (ZCS)). Note that it is unnecessary for both elements to be null at the moment of switching to eliminate losses.

Switch-based losses are the sum of conduction and switching losses. For a given loss budget (linked to the heat sink equipment), the smaller the switching losses are, the higher the switching frequency can be, allowing miniaturization of the passive components in the converter[1].

3.1.2. *Resonance*

Resonance is an interesting way of introducing current and voltage ripples so that they pass through zero at precise instants, while avoiding energy dissipation. Based on this principle, "inductance/capacitor"-type assemblies are particularly useful for soft switching in power electronic converters[2]. Consequently, *LC* circuits are always found in operating converters using soft switching; these circuits create resonance. For this reason, we often speak about resonant (or quasi-resonant) converters rather than soft switching converters.

This subject is complex and would fill a whole book on its own (e.g. [CHE 99] is devoted to this subject). In this chapter, we will focus on a single example, which is used wholly or partially in a wide range of applications: a DC–DC resonant converter, using two cascaded converters (as shown in Figure 3.1):

– a single-phase inverter connected to a resonant load;

– a diode rectifier (which is part of the resonant load).

1 Note, however, that switching losses in the power devices are not the only losses linked to the switching frequency. The skin effect in conductors and iron losses in magnetic circuits are intrinsically linked to their operating frequency. There is a limit (always of a thermal nature) to acceptable increases in the switching frequency (which is no longer systematically imposed by the switches).

2 Note the analogies between electrical currents and mechanical speed and between voltage and force. Using this analogy, an inductance is equivalent to a mass (or inertia), while a capacitor is equivalent to a spring.

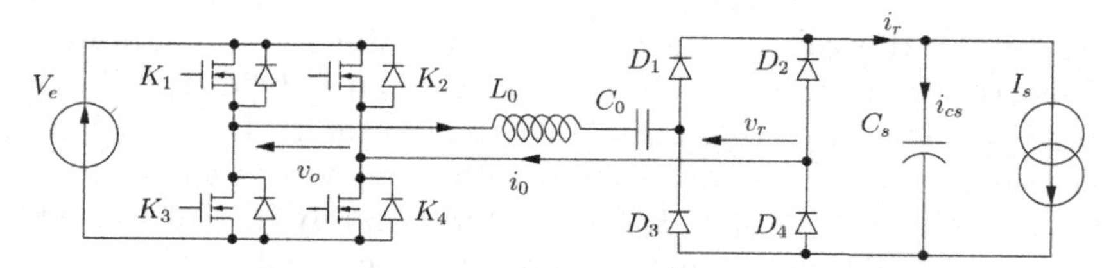

Figure 3.1. *DC–DC resonant converter*

3.2. Study of a resonant inverter

3.2.1. *Presentation*

We will begin by considering the inverter alone, replacing the diode rectifier (D_1, D_2, D_3, D_4) and its parallel load (I_s, C) by a resistance R_0. In these conditions, we may calculate the expression of the impedance of load Z_0:

$$Z_0 = R_0 + jL_0\omega + \frac{1}{jC_0\omega} \qquad [3.2]$$

Conventionally, equation [3.2] becomes:

$$Z_0 = \frac{1 + 2zj\omega/\omega_0 - (\omega/\omega_0)^2}{jC_0\omega} \qquad [3.3]$$

where the damping coefficient z and the natural angular frequency ω_0 are defined as follows:

$$z = \frac{R_0}{2}\sqrt{\frac{C_0}{L_0}} \qquad [3.4]$$

$$\omega_0 = \frac{1}{\sqrt{L_0 C_0}} \qquad [3.5]$$

Note that the inverter, made up of switches (K_1, K_2, K_3, K_4), is controlled using complementary signals for the switches in the same half-bridge: (K_1, K_3) on the one side and (K_2, K_4) on the other side, in order to avoid short circuits in the voltage source V_e. At this stage, no hypotheses concerning the effective control of the two half-bridges will be made, with the exception of the period. Thus, while considering the inverter's output voltage v_o with the period T, it can be decomposed as the sum of a (possibly infinite) set of simple oscillating functions, namely sines and cosines (or, equivalently, complex exponentials). The Fourier transform is a periodic function, often defined in terms of a Fourier series:

$$v_o(t) = a_0 + \sum_{k=1}^{\infty} a_k . \cos\left(\frac{2k\pi t}{T}\right) + b_k . \sin\left(\frac{2k\pi t}{T}\right) \qquad [3.6]$$

Evidently, the capacitance C_0 placed in series with the load means that no DC current will be able to circulate, and the average component a_0 which may be produced by the inverter is useless. The control process should therefore aim to make this component as small as possible to enable optimal use of the converter.

3.2.2. *Operating model*

A simple (but nonetheless satisfactory) model of the converter consists of approximating the voltage wave of v_o to the first harmonic, for two main reasons:

– as the load resonates with angular frequency ω_0, the converter should operate around this angular frequency, in order to maximize the power factor of the load and the amplitude of the supplied voltage[3];

– the damping coefficient z of the load is presumed to be high (although at this stage, this hypothesis is purely arbitrary).

3 It is clearly not desirable to base the resonance on a harmonic of the voltage v_o, which will generally have a lower amplitude than the fundamental, in the case where the inverter only switches twice per half-bridge and per period T.

In these conditions, only the fundamental component of the voltage wave v_o will be able to produce a current i_0 (sinusoidal) in the load. Hence, all of the harmonics of v_o may be omitted, so we may simply write:

$$v_o(t) = a_1 . \cos\left(\frac{2\pi t}{T}\right) + b_1 . \sin\left(\frac{2\pi t}{T}\right) \quad [3.7]$$

As the choice of a time origin is arbitrary, one can consider that:

$$v_o(t) = V_{omax} . \cos\left(\frac{2\pi t}{T}\right) \quad [3.8]$$

with a period T close to the natural period of the load:

$$T \simeq T_0 = \frac{2\pi}{\omega_0} = 2\pi\sqrt{L_0 C_0} \quad [3.9]$$

3.2.3. *Impact of the operating frequency*

In the simple case where the control signals of the two half-bridges are complementary with a 50% duty ratio, voltage v_o will be a square wave, with two plateaux at $\pm V_e$, each of width $T/2$. In a Fourier series decomposition of this type of signal, the amplitude of the fundamental can be written as:

$$V_{max} = \frac{4V_e}{\pi} \quad [3.10]$$

In the case where T is strictly equal to T_0, the current i_0 is perfectly in phase with the fundamental of the voltage. This means that at switching instants, the current is strictly null, thus the switching losses are negligible. However, the result is different if frequency $1/T$ varies:

– if the switching frequency is slightly higher, the load becomes inductive; therefore, the current presents a lag in relation to the voltage;

– if the switching frequency is slightly lower, the load becomes capacitive, and the current will be in advance of the voltage.

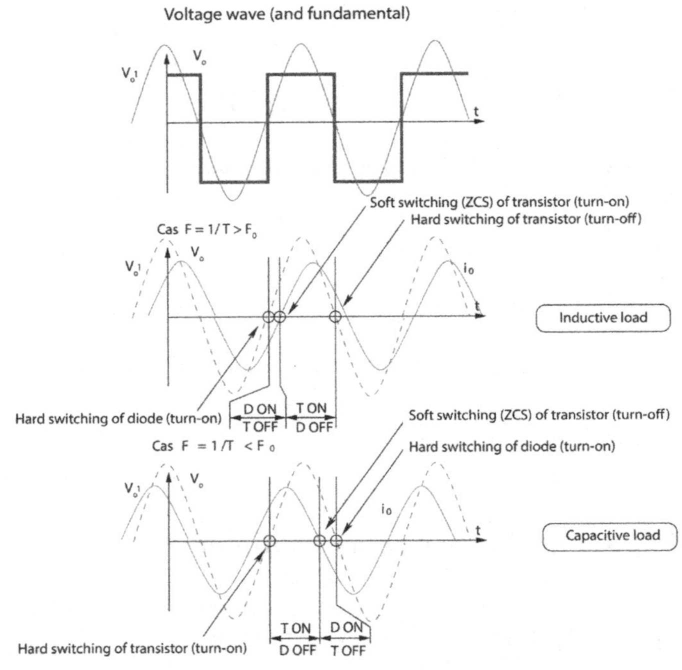

Figure 3.2. *Current / voltage waveforms at the first harmonic for output of a resonant inverter with* $F_0 = \frac{\omega_0}{2\pi}$

These two situations have an effect on switching losses, as shown in Figure 3.2. The information given on the nature of switching in the figure refers to K_1 (made up of a transistor and an antiparallel diode). If hard switching is not considered problematic for the diodes (as an initial hypothesis, spontaneous switching is presumed to be non-dissipative[4]), situation $F > F_0$ is clearly better for the transistor, insofar as the switch-on is soft, and minimizes the current peak effect

4 With the exception of losses which may occur when covering the potential barrier.

which occurs as a result of covering the potential barrier of the diode. In these conditions, the diode presents mediocre behavior in relation to this phenomenon, with a high stored charge Q_{rr}; the structure diode of a MOSFET may be used in this case.

3.2.4. *Power behaviors at variable frequency*

When operating using a variable frequency, using the hypothesis that the current i_0 absorbed by the load is quasi-sinusoidal, the complex current (for a phase voltage reference V_0) may be written as:

$$I_0 = \frac{V_{omax}}{R_0 + jL_0\omega + \frac{1}{jC_0\omega}} = \frac{jC_0\omega V_{omax}}{1 + jR_0C_0\omega - L_0C_0\omega^2} \quad [3.11]$$

which gives the following expression of a power P_0 dissipated in R_0:

$$P_0 = \frac{1}{2}R_0.\left|I_0\right|^2 = \frac{1}{2}\cdot\frac{R_0\left(C_0\omega V_{omax}\right)^2}{\left(1 - L_0C_0\omega^2\right)^2 + \left(R_0C_0\omega\right)^2} \quad [3.12]$$

The maximum value P_{0max} of this power, obtained for $\omega = \omega_0 = 1/\sqrt{L_0C_0}$, can then be noted as:

$$P_{0max} = \frac{V_{omax}^2}{2R_0} \quad [3.13]$$

It is now useful to reformulate the expression of P_0 to explicitly include P_{0max} and ω_0:

$$P_0 = \frac{1}{2R_0}\cdot\frac{\left(R_0C_0\omega V_{omax}\right)^2}{\left(1 - L_0C_0\omega^2\right)^2 + \left(R_0C_0\omega\right)^2} = \frac{P_{0max}}{\left(\frac{1-\left(\frac{\omega}{\omega_0}\right)^2}{R_0C_0\omega}\right)^2 + 1} \quad [3.14]$$

Note that $R_0 C_0 \omega = \frac{2z\omega}{\omega_0}$, leading to the expression:

$$P_0(\omega) = \frac{P_{0max}}{\frac{1}{4z^2}\left(\frac{\omega_0}{\omega} - \frac{\omega}{\omega_0}\right)^2 + 1} \quad [3.15]$$

The normalization of this power in relation to P_{0max} (introduction of a reduced power $y = P_0/P_{0max}$) and the introduction of a reduced angular frequency $x = \frac{\omega}{\omega_0}$ gives the following expression, of which traces are shown in Figure 3.3 (for different values of z, between 0.1 and 1):

$$y(x) = \frac{1}{\frac{1}{4z^2}\left(\frac{1}{x} - x\right)^2 + 1} \quad [3.16]$$

3.3. Study of the full converter

3.3.1. *Analysis of the diode rectifier*

Based on the hypothesis that operation can be assimilated to the first harmonic, the current circulating between the inverter and the rectifier (due to the presence of the series circuit $L_0 C_0$) is sinusoidal. In this situation, the rectifier no longer operates in a "classic" manner (as in Chapter 3 of Volume 2 [PAT 15b]) as a voltage rectifier/current inverter, but rather as a current rectifier/voltage inverter. When imposing a positive current i_0, we know that diodes D_1 and D_4 will necessarily be in a state of conduction; when the current is negative, diodes D_2 and D_3 will conduct. Using the hypothesis of a load circuit (I_s, C_s), such that the ripple of the output voltage v_s remains negligible in relation to its average value (denoted as V_s), we may state that v_r will be equal to $+V_s$ if i_0 is positive and $-V_s$ if i_0 is negative. As the current is sinusoidal with frequency $F = 1/T$ (to use the same notation as before), the time durations at $+V_s$ and $-V_s$ will be equal $(T/2)$. In these conditions, a square voltage v_r of amplitude V_s is obtained, with a null average value in phase with the sinusoidal current; the current has an amplitude denoted as I_{0max}. The rectifier load assembly may thus be considered as a pure resistance R_{eq} from the perspective of the resonant inverter (in the context of modeling around the first harmonic). We therefore need to determine the expression of this equivalent resistance. To this end, a link must be established between the current amplitude i_0 and the value of the current I_s absorbed by the load. To do this, note that our study concerns the converter in permanent mode; to guarantee periodic evolution of all quantities, the average current in the output capacitor C_s over a characteristic period T must be null. Hence, the average value of the rectified current $i_r(t) = |i_0(t)|$ must be equal to I_s. We have:

$$\langle i_r \rangle = \frac{1}{T}\int_0^T I_{0max}\left|\sin\left(\frac{2\pi t}{T}\right)\right| dt = \frac{2I_{0max}}{\pi} \quad [3.17]$$

Thus:

$$\frac{2I_{0max}}{\pi} = I_s \quad [3.18]$$

In fact, when the load is considered to be a current source, the amplitude I_{0max} of the current i_0 is independent of the applied voltage; hence, it is not possible to give an expression for the equivalent resistance. For this, we need to relate the current I_s to the voltage V_s by a load resistance $R_s = V_s/I_s$; under these conditions, we obtain:

$$V_s = \frac{2R_sI_{0max}}{\pi} \quad [3.19]$$

The amplitude V_{r1max} of the voltage fundamental applied to the rectifier input terminals is expressed as:

$$V_{r1max} = \frac{4V_s}{\pi} = \underbrace{\frac{8R_s}{\pi^2}}_{R_{eq}} \cdot I_{0max} \quad [3.20]$$

where the equivalent resistance R_{eq} of the rectifier + parallel load (R_s, C_s) naturally appears.

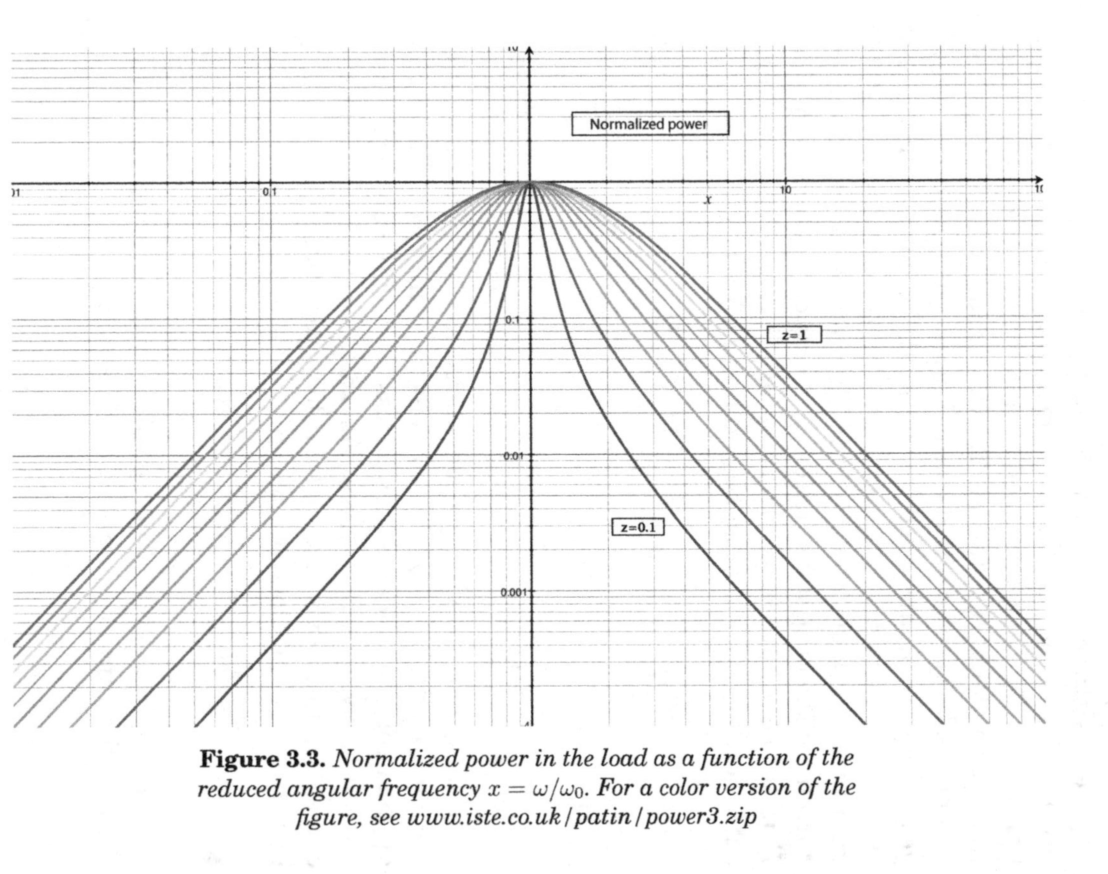

Figure 3.3. *Normalized power in the load as a function of the reduced angular frequency $x = \omega/\omega_0$. For a color version of the figure, see www.iste.co.uk/patin/power3.zip*

This equivalent resistance model shows that the results established for a resonant inverter with a series load R_0, L_0, C_0 are still applicable, even with a more complex load.

3.3.2. *Characteristics and control modes*

3.3.2.1. *Output characteristics*

Once the relationship between the load resistance R_s and the equivalent resistance R_{eq} has been established from the inverter's perspective, the relationship between the inverter's input voltage V_e and the rectifier's output voltage V_s may be established as a function of the way in which the inverter and the load are controlled. To achieve this goal, the operating conditions of the system must be clearly identified; note that, in any case, this study is based on modeling around the first harmonic, and the results obtained will always be dependent on the amplitude of the voltage v_o supplied by the inverter (denoted as V_{omax}). Note, simply, that the power dissipated in the equivalent resistance is always expressed in the form:

$$P_0(\omega) = \frac{P_{0max}}{\frac{1}{4z^2}\left(\frac{\omega_0}{\omega} - \frac{\omega}{\omega_0}\right)^2 + 1} \qquad [3.21]$$

with, in this case:

$$P_{0max} = \frac{\pi^2 V_{omax}^2}{8R_s} \qquad [3.22]$$

noting that this power can only be dissipated in the load R_s (all other elements are presumed to be ideal, and thus non-dissipative). Thus, we have:

$$\frac{V_s^2}{R_s} = \frac{\pi^2 V_{omax}^2}{8R_s} \cdot \frac{1}{\frac{1}{4z^2}\left(\frac{\omega_0}{\omega} - \frac{\omega}{\omega_0}\right)^2 + 1} \qquad [3.23]$$

Hence:

$$V_s = \frac{\pi V_{omax}}{2\sqrt{2}} \cdot \frac{1}{\sqrt{\frac{1}{4z^2}\left(\frac{\omega_0}{\omega} - \frac{\omega}{\omega_0}\right)^2 + 1}} \quad [3.24]$$

From this expression, it might seem that the converter's output voltage is independent of the load R_s; however, this parameter still has an effect via the damping coefficient z, expressed, in this case, as:

$$z = \frac{4R_s}{\pi^2}\sqrt{\frac{C_0}{L_0}} \quad [3.25]$$

Using these results, a converter control law may be proposed, with the aim of tuning the amplitude of V_{omax} or the angular frequency ω. These two solutions will be covered in the two following sections, and are known, respectively, as phase angle control and variable frequency control.

3.3.2.2. *Phase angle control mode*

In phase angle control mode, frequency $F = F_0 = \frac{\omega_0}{2\pi}$ is used. The waveforms of the controls are shown in Figure 3.4.

Using a phase angle ψ, the obtained voltage v_o returns to zero for periods of time which increase as ψ increases. This means that the amplitude of voltage fundamental component is expressed as:

$$V_{omax} = \frac{4V_e}{\pi}\cos\psi \quad [3.26]$$

Presuming that this component affects the current injected into the load, the objective in controlling V_{omax} has been reached. Next, to determine the action of the control input on the output voltage, the following equation is used:

$$V_s = \frac{\pi V_{omax}}{2\sqrt{2}} \cdot \frac{1}{\sqrt{\frac{1}{4z^2}\left(\frac{\omega_0}{\omega} - \frac{\omega}{\omega_0}\right)^2 + 1}} \quad [3.27]$$

replacing V_{omax} by its value and taking $\omega = \omega_0$. Hence:

$$V_s = \frac{2V_e}{\sqrt{2}} \cdot \cos\psi \qquad [3.28]$$

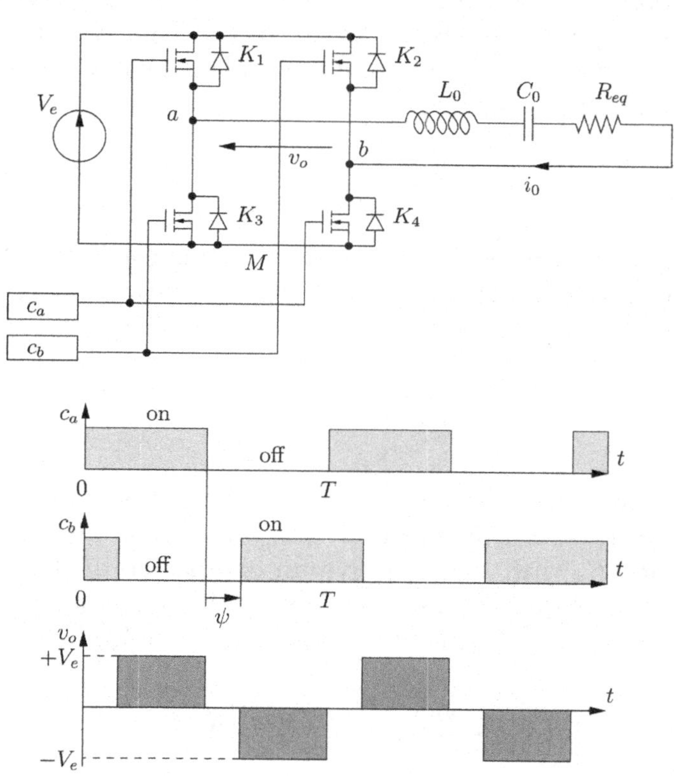

Figure 3.4. *Diagram and waveforms for a resonant inverter using phase angle control* ψ

This control scheme is interesting in that V_s may be directly controlled. Although the relationship is not linear, it is easy to regulate the output voltage; the performance of this strategy can be further improved by making the system linear in relation to the control input (ψ), integrating the "arccos" function into the controller.

3.3.2.3. *Variable frequency control mode*

In the case of variable frequency control, a square waveform $\pm V_e$ is also observed, generated very easily using two complementary control signals c_a and c_b produced by a voltage-controlled oscillator (VCO). In this case, we still have an amplitude V_{omax} for the fundamental component of voltage v_o, which is expressed as:

$$V_{omax} = \frac{4V_e}{\pi} \quad [3.29]$$

This time, however, the output voltage is controlled by the angular frequency ω:

$$V_s = \frac{\sqrt{2}V_e}{\sqrt{\frac{1}{4z^2}\left(\frac{\omega_0}{\omega} - \frac{\omega}{\omega_0}\right)^2 + 1}} \quad [3.30]$$

To reintroduce the load resistance into this expression, z is simply replaced by the relevant value. It is also possible to replace R_s with V_s/I_s to give an expression of the output voltage V_s in a more "traditional" form from a user perspective, i.e. as a function of the control input ω and the load current I_s:

$$\left(\frac{\pi^4 L_0 I_s^2}{64V_s^2 C_0}\left(\frac{\omega_0}{\omega} - \frac{\omega}{\omega_0}\right)^2 + 1\right).V_s^2 = 2V_e^2 \quad [3.31]$$

To simplify this expression, the following parameter may be used:

$$\lambda = \left(\frac{\omega_0}{\omega} - \frac{\omega}{\omega_0}\right)^2 \quad [3.32]$$

This gives the following equation:

$$\frac{\pi^4 L_0 \lambda I_s^2}{64C_0} + V_s^2 = 2V_e^2 \quad [3.33]$$

which may be reformulated as:

$$V_s^2 = 2V_e^2\left(1 - \frac{\pi^4 L_0 \lambda I_s^2}{128 C_0 V_e^2}\right) \qquad [3.34]$$

Introducing the normalized output voltage $y = \frac{V_s}{\sqrt{2}V_e}$ and the normalized output current $x = \frac{\pi^2}{8\sqrt{2}} \cdot \frac{I_s}{V_e} \cdot \sqrt{\frac{L_0}{C_0}}$, we obtain the equation:

$$y^2 = 1 - \lambda x^2 \qquad [3.35]$$

This is an ellipse equation, parameterized using λ. All of the obtained curves therefore take the value $y = 1$ for $x = 0$, whatever the value of λ. Moreover, these curves intersect with the x-axis ($y = 0$) at a value of x referred to as x_{cc} (corresponding to a short-circuit current I_{scc}), which is expressed as:

$$x_{cc} = \frac{1}{\sqrt{\lambda}} = \frac{1}{\frac{\omega_0}{\omega} - \frac{\omega}{\omega_0}} \qquad [3.36]$$

Note that there is a borderline case ($\omega = \omega_0$) where the value of x_{cc} tends toward infinity. It is easy to verify that at the resonant point, the obtained output voltage is independent of the load current as:

$$V_s = \sqrt{2}V_e \qquad [3.37]$$

A set of reduced characteristics $y(x)$ of the full converter (for different values of λ) is shown in Figure 3.5. Note that with this type of behavior, a converter controlled using variable frequency naturally produces an interesting operational safety property, in that it guarantees limitation of the load current, even allowing the output to short circuit.

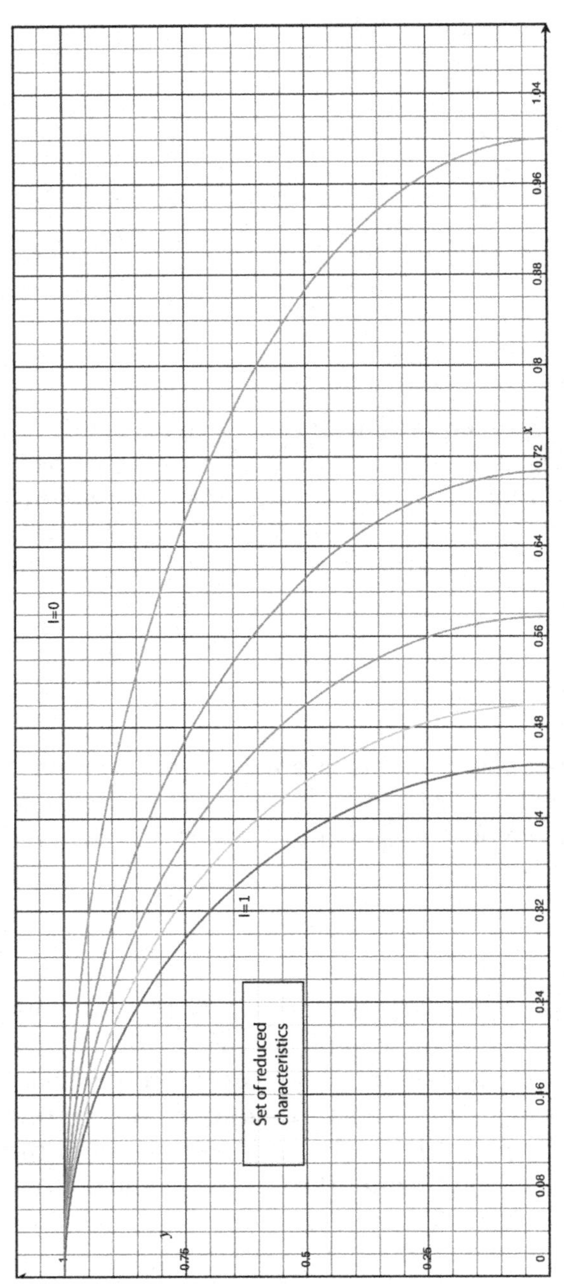

Figure 3.5. *Reduced output characteristics of a DC6DC resonant converter using variable frequency control. For a color version of the figure, see www.iste.co.uk/patin/power3.zip*

3.3.3. *Application to contactless power supplies*

A modification to the structure of the converter allows an isolated version to be created, replacing the coil by a transformer (see Figure 3.6). As we saw in Chapter 5 of Volume 1 [PAT 15a], this component may be modeled by an ideal coupler associated with a magnetizing inductance and a leakage inductance (as before, note that these elements may be localized either in the primary or the secondary of the transformer when modeling).

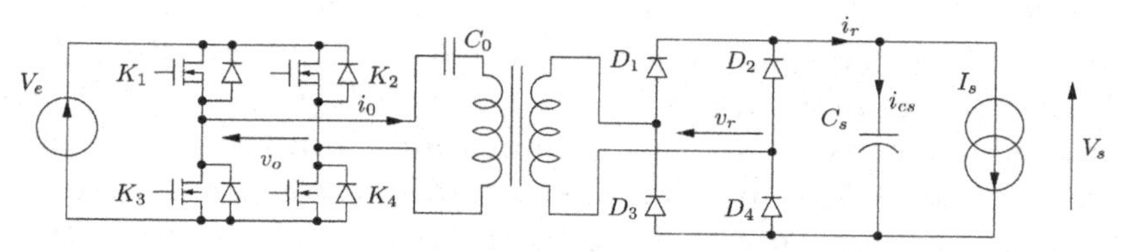

Figure 3.6. *Isolated DC–DC converter based on a resonant inverter and a rectifier*

This device clearly allows the introduction of the inductive element necessary for resonance, and also allows the impedance located in the secondary to be returned to the primary via a simple impedance transformation. It is easy to verify that the equivalent impedance Z_{eq} of an impedance Z viewed through a transformer with ratio m can be expressed as:

$$Z_{eq} = \frac{Z}{m^2} \qquad [3.38]$$

Thus, if a rectifier associated with a parallel RC charge is placed at the secondary of the transformer, and given the resistive nature of this assembly, as demonstrated in the previous sections, the same result is obtained, with a series RLC charging a resonant inverter. The previous results remain valid.

One interesting application using this type of structure is contactless energy transmission, where the primary and secondary windings of the transformer are connected magnetically by a two-part magnetic circuit, allowing the two elements to move freely in relation to each other. This type of device allows users, for example, to supply electrical power to an element located at the extremity of a robotic arm without limiting the movement of the arm itself by the use of wires; we also avoid the use of friction contact (slip ring and brush systems), which present drawbacks including wear and tear, sparks (which can be problematic when designing intrinsically safe equipment), etc. Note, however, that the magnetic circuit includes an air gap which goes against the desired objective; in this application, unlike a flyback transformer, where we like to store energy, we simply like to transfer energy directly between the primary and the secondary (a true transformer function, as in forward power supplies, with no air gap). In these conditions, the magnetizing inductance of the transformer will be relatively weak, leading to a high magnetizing current.

REMARK 3.1.– We have only considered the case of a resonant series circuit, but parallel resonant circuits may also be used. Although the two solutions are equivalent in theory, they are dimensioned differently, and solutions will be selected based on their advantages for specific applications. Furthermore, note that the association of capacitors with the inductive elements placed in resonant circuits may be addressed in terms of an increase in the power factor, as in a classic electrical engineering problem using a 50 Hz network.

4

Converter Modeling for Control

4.1. Principles

The aim of this chapter is to present a method of modeling switch-mode power supplies which allows establishing their transfer functions. On this basis, a classical control design methodology can be applied in order to achieve an output voltage regulation for instance. To do this, the average switch model, proposed by Middlebrook in 1976 at the IEEE Power Electronics Specialists Conference (PESC) and published [MID 77] in 1977, is used. For this, the switches in a switching cell (a transistor and a diode) are replaced by two equivalent sources (voltage and current) linked to the state variables (voltage at the capacitor terminals or current in an inductance) and to the constant sources in the system. An equivalent means of carrying out this modeling is to apply a state formalism, which describes the system in question in the form of a first-degree differential (vector) equation:

$$\frac{d\mathrm{x}}{dt} = A.\mathrm{x} + B.\mathrm{u} \qquad [4.1]$$

where x is the system state vector and u is the input. Note that in these conditions the system is linear; further note that all switch-mode power supplies are hybrid linear systems (switching from one linear equation to another in a period of time determined by the controller). Using the equations obtained for each time interval, an average state equation can be established in order to obtain the desired transfer functions for the system.

4.2. Continuous conduction modeling

This section is devoted to non-isolated converters, examining the case of the buck, boost and buck–boost converters; this approach is easily applied to the corresponding isolated converters (except for the boost, where no isolated version exists). Generally speaking, all of the considered quantities $x(t)$ (voltages and currents) take the form:

$$x(t) = \langle x \rangle + \tilde{x}(t) \qquad [4.2]$$

where $\tilde{x}(t)$ is the fluctuating component (with an average value of zero) of $x(t)$. For reasons of simplicity, $\langle x \rangle$ (for the average value of $x(t)$) will be replaced by x_0. Generally speaking, the following sections will only consider the converters using small signals, and therefore concern only the fluctuating components. To do this, linearization is required around an operating point in the converter equations.

The equations for a switch-mode power supply can easily be placed into state form by identifying the elements carrying each state variable: these are all energy storage elements, specifically:

– inductances: the current traveling through an inductance is a state variable;

– capacitors: the voltage at the terminals of a capacitor (ideal) is a state variable.

Given that the converters in question involve both an inductance and a capacitor (in the case of non-isolated converters), second-order systems must be used. The state vector of the switch-mode power supplies studied in the next few sections is, therefore, always of the form $\mathrm{x} = \begin{pmatrix} i_L & v_s \end{pmatrix}^t$, where v_s is the converter output voltage (as the capacitor is always placed at the converter output point – buck, boost and buck–boost).

4.2.1. *The buck converter*

The state equation formulation of the model of a buck converter consists of associating the equations related to the inductance and the capacitor in a vector-based system:

$$\begin{cases} \frac{di_L}{dt} = \frac{1}{L}\left(k.v_e - v_s\right) \\ \frac{dv_s}{dt} = \frac{1}{C}\left(i_L - \frac{v_s}{R}\right) \end{cases} \qquad [4.3]$$

where k is the instantaneous control of the transistor (0 when OFF and 1 when ON). Note that the average of these equations at the level of the switching period T_{sw} simply consists of replacing k by the duty ratio $\alpha = \langle k \rangle$ in these instantaneous equations in order to obtain the average equations (average model):

$$\begin{cases} \frac{di_L}{dt} = \frac{1}{L}\left(\alpha.v_e - v_s\right) \\ \frac{dv_s}{dt} = \frac{1}{C}\left(i_L - \frac{v_s}{R}\right) \end{cases} \qquad [4.4]$$

Thus, the state equation becomes:

$$\frac{d\mathbf{x}}{dt} = A.\mathrm{x} + B.\mathrm{u} \qquad [4.5]$$

with:

$$A = \begin{pmatrix} 0 & -\frac{1}{L} \\ \frac{1}{C} & -\frac{1}{RC} \end{pmatrix} \qquad [4.6]$$

The stability of this system can be easily analyzed by calculating the roots of the characteristic polynomial; however, in this case, we simply wish to establish a usable model for control purposes. The control signal appears in the second term of the state equation, B.u; in our case, the term cannot be directly identified with a linear model, as we have a vector (denoted as w) expressed as:

$$\mathbf{w} = \begin{pmatrix} \frac{\alpha . v_e}{L} \\ 0 \end{pmatrix} \qquad [4.7]$$

The first coefficient is clearly nonlinear, given that the input voltage v_e may fluctuate. In this case, w may be linearized as follows:

$$\mathbf{w} = \underbrace{\begin{pmatrix} \alpha_0 . V_{e0} \\ 0 \end{pmatrix}}_{\mathbf{w}_0} + \begin{pmatrix} \alpha_0 . \tilde{v}_e \\ 0 \end{pmatrix} + \begin{pmatrix} \tilde{\alpha} . V_{e0} \\ 0 \end{pmatrix} \qquad [4.8]$$

From the state equation, the static operation point $\mathbf{x}_0$ is easy to identify:

$$0 = A.\mathbf{x}_0 + \mathbf{w}_0 \implies \mathbf{x}_0 = -A^{-1}.\mathbf{w}_0 \qquad [4.9]$$

i.e.:

$$\begin{cases} I_{L0} = \frac{\alpha_0 . V_{e0}}{R} \\ V_{s0} = \alpha_0 . V_{e0} \end{cases} \qquad [4.10]$$

The equation form of the variations, denoted as $\tilde{\mathrm{x}}$ ("small signal" model), of state x is obtained as follows:

$$\frac{d\tilde{\mathbf{x}}}{dt} = A.\tilde{\mathbf{x}} + \frac{1}{L} \cdot \begin{pmatrix} \alpha_0 \\ 0 \end{pmatrix} .\tilde{v}_e + \frac{1}{L} \cdot \begin{pmatrix} V_{e0} \\ 0 \end{pmatrix} .\tilde{\alpha} \qquad [4.11]$$

The transfer functions of the system can be easily deduced, for example, taking $\tilde{v}_e = 0$ to obtain the relationships linking

$\tilde{\alpha}$ to components $\tilde{i}_L$ and $\tilde{v}_s$ of the state vector. To do this, note that in the Laplace domain, $\frac{d\tilde{\mathbf{x}}}{dt}$ is written as $s\tilde{\mathbf{X}}$. Furthermore, note that:

$$i_L = \underbrace{\begin{pmatrix} 1\,0 \end{pmatrix}}_{C_1}.\tilde{\mathbf{X}} \qquad [4.12]$$

and:

$$v_s = \underbrace{\begin{pmatrix} 0\,1 \end{pmatrix}}_{C_2}.\tilde{\mathbf{X}} \qquad [4.13]$$

hence:

$$\begin{aligned} H_i(s) &= \frac{\tilde{I}_L(s)}{\tilde{\alpha}(s)} = C_1.(s\mathbb{I}_2 - A)^{-1}.\begin{pmatrix} V_{e0}/L \\ 0 \end{pmatrix} \\ &= \frac{V_{e0}}{R} \cdot \frac{1 + RCs}{1 + \frac{Ls}{R} + LCs^2} \end{aligned} \qquad [4.14]$$

and:

$$\begin{aligned} H_v(s) &= \frac{\tilde{V}_s(s)}{\tilde{\alpha}(s)} = C_2.(s\mathbb{I}_2 - A)^{-1}.\begin{pmatrix} V_{e0}/L \\ 0 \end{pmatrix} \\ &= \frac{V_{e0}}{1 + \frac{Ls}{R} + LCs^2} \end{aligned} \qquad [4.15]$$

In the same way, a relationship linking $\tilde{v}_s$ and $\tilde{v}_e$ can be established, taking:

$$\begin{aligned} H_{es}(s) &= \frac{\tilde{V}_s(s)}{\tilde{V}_e(s)} = C_2.(s\mathbb{I}_2 - A)^{-1}.\begin{pmatrix} \alpha_0/L \\ 0 \end{pmatrix} \\ &= \frac{\alpha_0}{1 + \frac{Ls}{R} + LCs^2} \end{aligned} \qquad [4.16]$$

4.2.2. *The buck–boost converter*

The same approach may be used for the buck–boost converter, but the average equations cannot be written directly, as:

$$\frac{di_L}{dt} = \frac{v_e}{L} \qquad [4.17]$$

during the first part (phase 1) of the switching period ($0 \leq t \leq \alpha.T_{sw}$) and:

$$\frac{di_L}{dt} = -\frac{v_s}{L} \qquad [4.18]$$

for the remaining of the period ($\alpha.T_{sw} \leq t \leq T_{sw}$, i.e. phase 2). Consequently, the average equation relating to i_L becomes:

$$\frac{di_L}{dt} = \frac{\alpha.v_e}{L} - \frac{(1-\alpha).v_s}{L} \qquad [4.19]$$

We can then note that, during phase 1:

$$\frac{dv_s}{dt} = -\frac{v_s}{RC} \qquad [4.20]$$

and, during phase 2:

$$\frac{dv_s}{dt} = \frac{1}{C}\left(i_L - \frac{v_s}{R}\right) \qquad [4.21]$$

giving an average equation of the form:

$$\frac{dv_s}{dt} = -\frac{\alpha.v_s}{RC} + \frac{1-\alpha}{C}\left(i_L - \frac{v_s}{R}\right) \qquad [4.22]$$

The equations of the state system obtained in this way are nonlinear, and must be linearized around an operating point, by considering the following:

$$\begin{cases} v_e = V_{e0} + \tilde{v}_e \\ \alpha = \alpha_0 + \tilde{\alpha} \\ i_L = I_{L0} + \tilde{i}_L \\ v_s = V_{s0} + \tilde{v}_s \end{cases} \quad [4.23]$$

The equations relating to the bias point can then be established:

$$0 = \frac{\alpha_0.v_{e0}}{L} - \frac{(1-\alpha_0).V_{s0}}{L} \quad [4.24]$$

hence:

$$V_{s0} = \frac{\alpha_0.V_{e0}}{1-\alpha_0} \quad [4.25]$$

and:

$$0 = -\frac{\alpha_0.V_{s0}}{RC} + \frac{1-\alpha_0}{C}\left(I_{L0} - \frac{V_{s0}}{R}\right) \quad [4.26]$$

giving the result:

$$I_{L0} = \frac{V_{s0}}{(1-\alpha_0).R} \quad [4.27]$$

V_{s0} may then be replaced by its expression as a function of the system input quantities (α_0 and V_{e0}) and the load R to obtain:

$$I_{L0} = \frac{\alpha_0.V_{s0}}{(1-\alpha_0)^2.R} \quad [4.28]$$

By linearizing the state equations, it is possible to write:

$$\frac{d\tilde{i}_L}{dt} = -\frac{(1-\alpha_0).\tilde{v}_s}{L} + \frac{V_{e0}.\tilde{\alpha}}{(1-\alpha_0).L} + \frac{\alpha_0.\tilde{v}_e}{L} \quad [4.29]$$

and:

$$\frac{d\tilde{v}_s}{dt} = \frac{1-\alpha_0}{C} \cdot \tilde{i}_L - \frac{1}{RC} \cdot \tilde{v}_s - \frac{\alpha_0 . V_{e0}}{(1-\alpha_0)^2 . RC} \cdot \tilde{\alpha} \qquad [4.30]$$

This gives a linearized state system of the form:

$$\frac{d\tilde{\mathbf{x}}}{dt} = \underbrace{\begin{pmatrix} 0 & -\frac{1-\alpha_0}{L} \\ \frac{1-\alpha_0}{C} & -\frac{1}{RC} \end{pmatrix}}_{A} .\tilde{\mathbf{x}} + \underbrace{\begin{pmatrix} \frac{\alpha_0}{L} \\ 0 \end{pmatrix}}_{B_1} .\tilde{v}_e + \underbrace{\begin{pmatrix} \frac{V_{e0}}{(1-\alpha_0).L} \\ -\frac{\alpha_0 . V_{e0}}{(1-\alpha_0)^2 . RC} \end{pmatrix}}_{B_2} .\tilde{\alpha} \qquad [4.31]$$

This system can then be used (as for the buck converter) to deduce the following transfer functions:

$$H_v(s) = \frac{\tilde{V}_s(s)}{\tilde{\alpha}(s)} = C_2 . (s\mathbb{I}_2 - A)^{-1} . B_2$$

$$= \frac{V_{e0}}{(1-\alpha_0)^2} \cdot \frac{1 - \frac{\alpha_0 L s}{(1-\alpha_0)^2 . R}}{1 + \frac{Ls}{(1-\alpha_0)^2 . R} + \frac{LCs^2}{(1-\alpha_0)^2}} \qquad [4.32]$$

and:

$$H_{es}(p) = \frac{\tilde{V}_s(s)}{\tilde{V}_e(s)} = C_2 . (s\mathbb{I}_2 - A)^{-1} . B_1$$

$$= \frac{\alpha_0}{1-\alpha_0} \cdot \frac{1}{1 + \frac{Ls}{(1-\alpha_0)^2 . R} + \frac{LCs^2}{(1-\alpha_0)^2}} \qquad [4.33]$$

Note that matrices C_1 and C_2 are the same as for the buck converter, in that the configuration of the state vector remains the same (with i_L as the first component and v_s as the second component). The specificity of this chopper lies in the fact that it is a non-minimum phase system, as seen in the transfer function $H_v(s)$ where the real positive part has a value of zero. With this transfer function, the increase in the duty ratio leads, temporarily, to a reduction in the output voltage, and then to an increase, in accordance with the static behavior, for which the output voltage is written as $\frac{\alpha_0 . V_{e0}}{1-\alpha_0}$.

4.2.3. *The boost converter*

Using the same approach as for the buck and buck–boost converters, the following transfer functions are obtained:

$$H_v(s) = \frac{\tilde{V}_s(s)}{\tilde{\alpha}(s)} = \frac{V_{e0}}{1-\alpha_0} \cdot \frac{1 - \frac{Ls}{(1-\alpha_0)^2.R}}{1 + \frac{Ls}{(1-\alpha_0)^2.R} + \frac{LCs^2}{(1-\alpha_0)^2}} \qquad [4.34]$$

and:

$$H_{es}(s) = \frac{\tilde{V}_s(s)}{\tilde{V}_e(s)} = \frac{1}{1-\alpha_0} \cdot \frac{1}{1 + \frac{Ls}{(1-\alpha_0)^2.R} + \frac{LCs^2}{(1-\alpha_0)^2}} \qquad [4.35]$$

4.3. Discontinuous conduction modeling

Static converters can be shown to be more stable in discontinuous conduction than in continuous conduction, resulting in first-order transfer functions. These functions are shown in Table 4.1, using results taken from [FER 02]. Note that the converter output voltage (in static mode) is no longer a function of the duty ratio α_0 and the input voltage V_{e0}, but also of the load. All of the transfer functions include a reduced static output voltage (i.e. around an operating point) $y_0 = V_{s0}/V_{e0}$. To simplify the equations, given that these parameters exist for all of the converters in question, we will use coefficient K and a characteristic angular frequency ω_c, defined as follows:

$$\begin{cases} K = \frac{2LF_{sw}}{R} \\ \omega_c = \frac{1}{RC} \cdot \frac{2y_0 - 1}{y_0 - 1} \end{cases} \qquad [4.36]$$

4.4. PWM control modeling and global modeling for control

A model of pulse width modulation (PWM) control needs to be established before designing a control strategy for the converter output voltage, as we develop analog or pseudo-analog control[1] to produce binary or digital-type control of the transistor in the electrical power converter. The interface between the "analog" control of the system and the digital control of the transistor is provided by a PWM controller, allowing control of the duty ratio α of the control output[2].

Converters	H_v	H_{es}
Buck	$\frac{V_{e0}\,(1-y_0)}{2-y_0}\cdot\sqrt{\frac{1-y_0}{K}}\cdot\frac{1}{1+\frac{s}{\omega_c}}$	$\frac{y_0}{1+\frac{s}{\omega_c}}$
Buck–boost	$\frac{V_{e0}}{\sqrt{K}}\cdot\frac{1}{1+\frac{RCp}{2}}$	$\frac{y_0}{1+\frac{RCs}{2}}$
Boost	$\frac{2V_{e0}}{2y_0-1}\cdot\sqrt{\frac{y_0\,(y_0-1)}{K}}\cdot\frac{1}{1+\frac{p}{\omega_c}}$	$\frac{y_0}{1+\frac{s}{\omega_c}}$

Table 4.1. *Transfer functions for switch-mode power supplies in discontinuous conduction mode*

Note that the creation of a PWM controller in an analog form consists of comparing a reference value V_{ref} which is constant (or varies slowly in relation to the length of the switching period) with a triangular carrier V_p varying, for example, between 0 and a voltage $V_{\max}$. Within this variation range, a value of 0 (OFF) or 1 (ON) is associated with the output according to the following conditions:

– control signal = 1 if $V_{\text{ref}} > V_p$;

– control signal = 0 if $V_{\text{ref}} < V_p$.

1 Using a digital controller, for example, the digital signal processor (DSP).

2 In this case, we will only consider fixed frequency control, using a converter with a controllable duty ratio. We will not consider hysteresis or "current bracket"-type controls, which are also widely used in this context, and form part of sliding mode control approaches in automatics.

Note that the control signal then presents a duty ratio α with an expression which evolves in a linear manner with V_{ref}:

$$\alpha = \frac{V_{\mathrm{ref}}}{V_{\max}} \qquad [4.37]$$

The modulator (PWM controller) is, therefore, equivalent to a pure gain ($K_{PWM} = 1/V_{\max}$) linking the duty ratio α to the reference signal V_{ref}. Furthermore, note that, somewhat surprisingly, this control approach does not introduce a lag, as we might expect. PWM control only results in a delay in the case of a digital implementation, where the reference is sampled (and mathematically blocked with a zero-order hold). In this case, our model must include a transfer function, $B_0(p)$, of the form:

$$B_0(s) = \frac{1 - e^{-T_{sw}s}}{s} \qquad [4.38]$$

With the delay, this may be approximated to:

$$H_r(s) = e^{-\frac{T_{sw}s}{2}} \qquad [4.39]$$

4.5. General block diagram of a voltage-regulated power supply

Based on these findings, a full block diagram of closed-loop regulation of the output voltage of a switch-mode power supply, such as the one shown in Figure 4.1, can be established. The physical elements of the system (converter, PWM controller and voltage capacitor) must all be characterized in order to determine the transfer function for a regulator $C(s)$ placed in the loop. This is a classic problem in control, where a given response time needs to be obtained while limiting oscillation, and avoiding the disturbances induced by the load and/or by fluctuations in the input voltage. Moreover, it is necessary to ensure that the regulator is robust in relation to variations in the system parameters, which may fluctuate, notably as a function of temperature.

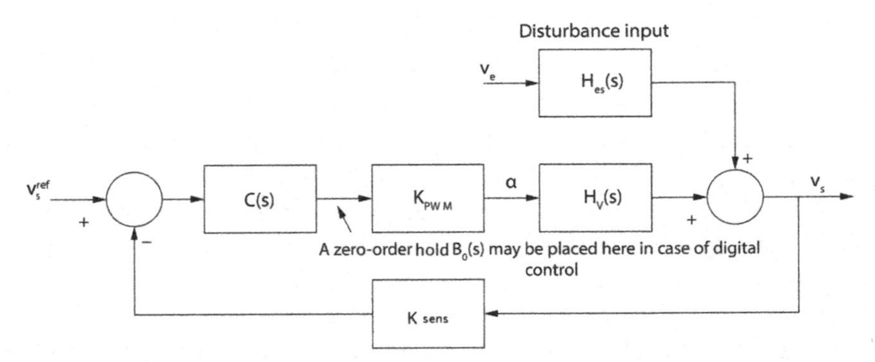

Figure 4.1. *Generic block diagram of output voltage regulation in a switch-mode power supply*

5

Case Study – the Flyback Power Supply

5.1. Specification

In this chapter, we wish to design an "isolated" type switch-mode power supply, intended to supply a 15 W load using 15 V (i.e. with an output current of 1 A) from a 5 V source. In this context, the output voltage should have a quality value of $\pm 1\%$ of the nominal voltage. To achieve this, a converter, operating at a frequency considerably higher than the audible spectrum ($F_{sw} = 50\,\text{kHz}$) and allowing miniaturization of the passive components, is used. These starting hypotheses mean that a Flyback-type power supply (see Figure 5.1) is a good choice, as the converter:

– includes galvanic isolation between the input and the output;

– has a boost operating capability (output voltage higher than the input voltage) independently of the turns ratio of the coupled inductances;

– is simple (in comparison with a Forward converter) and suitable for use at the low power levels required.

Based on this choice, we first need to establish the rated voltage and current of the required switches (transistor and diode), then design the coupled inductances (selection of a core and calculation of the number of turns required for the primary and secondary windings). Once these inductances have been created, measurements need to be carried out in order to evaluate the quality of the magnetic coupling and the value of the leakage inductance localized at the primary winding in order to dimension a snubber. Finally, a closed-loop control circuit will be implemented to ensure that the output voltage remains at 15 V in spite of potential variations of consumption in the load. Figure 5.2 shows a full diagram of a Flyback power supply designed using this method (with the exception of the digital control circuit, which, in this case, is a Microchip dsPIC33FJ06GS101 microcontroller – including a DSP core designed for the control of switch-mode power supplies).

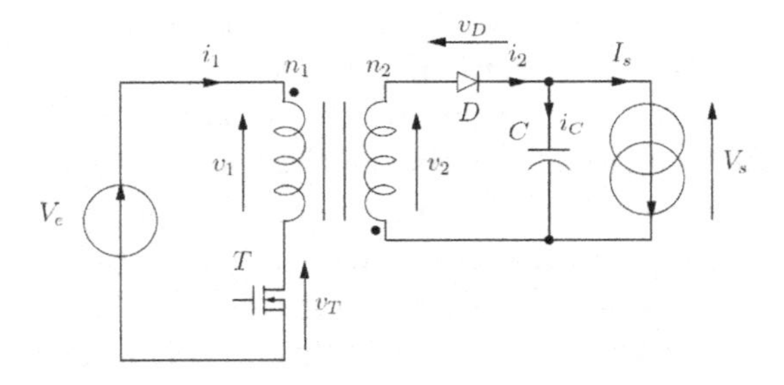

Figure 5.1. *Simplified diagram of a Flyback power supply*

5.2. Dimensioning switches

In Chapter 1, the constraints applicable to switches in a Buck-Boost converter were established; we also showed that these results can be transposed to the Flyback power supply (see Table 5.1) from which this converter chopper is derived, via the turns ratio $m = n_2/n_1$.

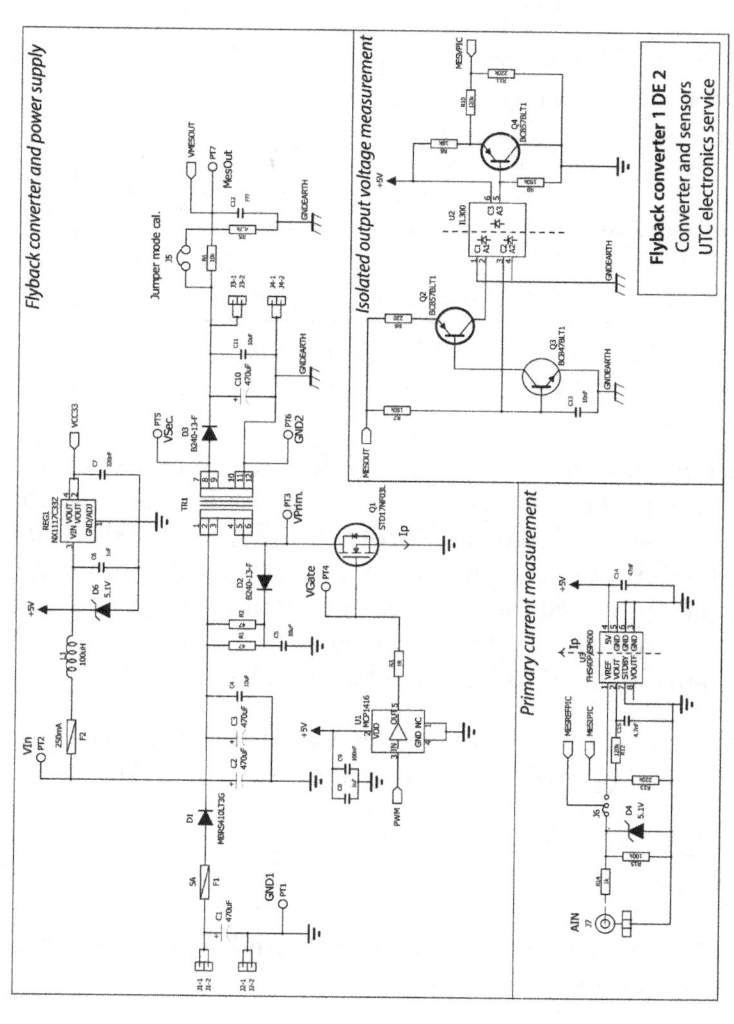

Figure 5.2. *Diagram of a switch-mode Flyback power supply (created using Mentor Graphics PADS®)*

Quantities	*Values*
Max. transistor voltage $V_{T\max}$	$V_e + \frac{V_s}{m}$
Max. inverse diode voltage $V_{d\max}$	$-(m.V_e + V_s)$
Current ripple in primary inductance Δi_{L1}	$\frac{\alpha.V_e}{L_1.F_{sw}}$
Average output voltage $\langle V_s \rangle$	$\frac{\alpha.m.V_e}{1-\alpha}$
Output voltage ripple Δv_s	$\frac{\alpha.I_s}{C.F_{sw}}$
Max. currents (transistor and diode) (as there are two)	$i_{T\max} = \frac{\alpha.V_e}{L_1.F_{sw}}$ and $i_{D\max} = \frac{\alpha.V_s}{m.L_1.F_{sw}} = \frac{\alpha.V_s}{M.F_{sw}}$
Average current in the transistor $\langle I_T \rangle$	$\frac{\alpha^2.V_e}{2L_1.F_{sw}}$
Average current in the diode $\langle I_d \rangle$	$\frac{\alpha(1-\alpha).V_e}{2M.F_{sw}}$

Table 5.1. *Summary of critical conduction in the Flyback power supply*

Semiconductor design is closely linked to the turns ratio m at this level. The turns ratio is also involved in the relationship between input and output voltages. An arbitrary choice therefore needs to be made, as there are an "infinite" number of (m, α) couples which will lead to the desired solution. In this case, a duty ratio close to 1/2 ($\alpha_0 = 0.5$) will be used. This leads us to choose a turns ratio $m = 3$.

On this basis, the voltage values which the transistor and diode will need to withstand can be determined as follows:

$$\begin{cases} V_{T\max} = 5 + \frac{15}{3} = 10\,\text{V} \\ V_{D\max} = -(15 + 15) = -30\,\text{V} \end{cases} \quad [5.1]$$

Security coefficients must be included when choosing voltage calibers for the two components.

In terms of current, as the nominal output current is 1 A, the average current in the diode will be equal to this value. With a duty ratio of 0.5 and in the context of critical conduction, we note that the RMS value of this current is easy to calculate, with a high frequency ripple of amplitude $I_{D\max}$ such that:

$$\langle I_D \rangle = \frac{1}{T_d}\int_0^{0.5T_d} \frac{I_{D\max} t}{0.5T_d} dt = \frac{I_{D\max}}{4} \quad [5.2]$$

This gives a maximum current of 4 A, from which the RMS value can be calculated:

$$I_{D\mathrm{RMS}} = \sqrt{\frac{1}{T_d}\int_0^{0,5T_d} \left(\frac{I_{D\max} t}{0.5T_d}\right)^2 dt} = \sqrt{\frac{I_{D\max}^2}{2}\int_0^1 u^2 du}$$

$$= \frac{I_{D\max}}{\sqrt{6}} = 1.63\,\mathrm{A} \quad [5.3]$$

As the waveform of the current in the transistor is exactly symmetrical to that of the current in the diode within a switching period, the results take the same form; we simply note that, by using the identity between the converter input and output powers (unitary efficiency hypothesis for switch dimensioning), the average input current is equal to 3 A. From this, a maximum current value $I_{T\max}$ of 12 A is deduced, hence:

$$I_{T\mathrm{RMS}} \simeq 4.90\,\mathrm{A} \quad [5.4]$$

In summary, we require:

– a transistor (MOSFET) able to withstand minimum RMS values of 10 V and 4.90 A;

– a diode (Schottky, for example) able to withstand an inverse voltage of at least 30 V and an average current of 1 A[1].

1 The RMS current is less significant for a diode, as the voltage dropoff in the ON state is generally almost constant as a function of the current. Nevertheless, the RMS current may be used, if we take account of the dynamic resistance in the ON state (which is generally very low).

For example, we may use the following components, chosen notably for their wide availability from the usual sources (Farnell, Radiospares, Digikey, Mouser, etc.):

– the MOSFET transistor (ST Microelectronics) STD17NF03 in a SMD DPAK package with a rated voltage and current of 30V/17A. CHIFFRES MANQUANTES;

– a Schottky diode (Diodes Incorporated) B240-13-F with a rated voltage and current of 40 V/2 A[2].

Figures 5.3 and 5.4 show extracts from the datasheets relating to the two components.

5.3. Calculation of passive components

5.3.1. *Output capacitors*

The ripple of the output voltage Δv_s is dependent on the load current I_s, the duty ratio α, the (overall) capacitance C of the output capacitor and the switching frequency F_d at which the converter operates. In this case, we have:

– $\Delta v_s = 0.02 \times 15 = 0.3\,\text{V}$, from the specification established at the beginning of the study;

– $I_s = 1\,\text{A}$, again from the specification;

– F_d was fixed at $50\,\text{kHz}$ in the same way, for miniaturization purposes (acoustical aspects cease to be problematic above $20\,\text{kHz}$);

– the operation of the power supply has been fixed arbitrarily at a duty ratio α of around 0.5.

2 This diode is suitable in terms of the voltage caliber, but it is considerably overdimensioned in terms of the current caliber. All of the Schottky diodes contained in the IRF catalog are overdimensioned; although diodes with lower calibers exist, this particular option was selected as it is a SMD component, and can easily be obtained from the same supplier as the transistors (component acquisition can itself be problematic).

STD17NF03L

N-CHANNEL 30V - 0.038Ω - 17A - DPAK/IPAK
STripFET™ POWER MOSFET

TYPE	V_{DSS}	$R_{DS(on)}$	I_D
STD17NF03L	30V	$<0.05\Omega$	17A

- TYPICAL R_{DS}(on) = 0.038Ω
- EXCEPTIONAL dv/dt CAPABILITY
- APPLICATION ORIENTED CHARACTERIZATION
- ADD SUFFIX "T4" FOR ORDERING IN TAPE & REEL
- ADD SUFFIX "-1" FOR ORDERING IN IPAK VERSION

3
2
1
IPAK

3
1
DPAK

DESCRIPTION

This Power Mosfet is the latest development of STMicroelectronics unique "Single Feature Size™" strip-based process. The resulting transistor shows extremely high packing density for low on-resistance, rugged avalance characteristics and less critical alignment steps therefore a remarkable manufacturing reproducibility.

INTERNAL SCHEMATIC DIAGRAM

D(2)
G(1)
S(3)
SC06140

APPLICATIONS
- DC-DC & DC-AC CONVERTERS
- MOTOR CONTROL, AUDIO AMPLIFIERS
- SOLENOID AND RELAY DRIVERS
- AUTOMOTIVE ENVIRONMENT

ABSOLUTE MAXIMUM RATINGS

Symbol	Parameter	Value	Unit
V_{DS}	Drain-source Voltage (V_{GS} = 0)	30	V
V_{DGR}	Drain-gate Voltage (R_{GS} = 20 kΩ)	30	V
V_{GS}	Gate- source Voltage	±20	V
I_D	Drain Current (continuos) at T_C = 25°C	17	A
I_D	Drain Current (continuos) at T_C = 100°C	12	A
I_{DM} (•)	Drain Current (pulsed)	68	A
P_{TOT}	Total Dissipation at T_C = 25°C	20	W
	Derating Factor	0.13	W/°C
dv/dt (1)	Peak Diode Recovery voltage slope	6	V/ns
E_{AS} (2)	Single Pulse Avalanche Energy	200	mJ
T_{stg}	Storage Temperature	–65 to 175	°C
T_j	Max. Operating Junction Temperature	175	°C

(•) Pulse width limited by safe operating area

(1) $I_{SD} \leq 17A$, di/dt ≤300A/μs, $V_{DD} \leq V_{(BR)DSS}$, $T_j \leq T_{JMAX}$.
(2) Starting T_j=25°C, I_D=11A, V_{DD}=15V

Aug 2000 1/9

Figure 5.3. *Extract from transistor documentation (ST Microelectronics)*

B220/A - B260/A

2.0A SURFACE MOUNT SCHOTTKY BARRIER RECTIFIER

Features

- Guard Ring Die Construction for Transient Protection
- Ideally Suited for Automated Assembly
- Low Power Loss, High Efficiency
- Surge Overload Rating to 50A Peak
- For Use in Low Voltage, High Frequency Inverters, Free Wheeling, and Polarity Protection Application
- High Temperature Soldering: 260°C/10 Second at Terminal
- **Lead Free Finish/RoHS Compliant (Note 1)**
- **Green Molding Compound (No Halogen and Antimony) (Note 2)**

Mechanical Data

- Case: SMA/SMB
- Case Material: Molded Plastic. UL Flammability Classification Rating 94V-0
- Moisture Sensitivity: Level 1 per J-STD-020
- Terminals: Lead Free Plating (Matte Tin Finish). Solderable per MIL-STD-202, Method 208 (e3)
- Polarity: Cathode Band or Cathode Notch
- Weight: SMA 0.064 grams (Approximate)
  SMB 0.093 grams (Approximate)

Top View

Bottom View

Ordering Information (Note 3)

Part Number	Case	Packaging
B2xxA-13-F	SMA	5000/Tape & Reel
B2xx-13-F	SMB	3000/Tape & Reel

* x = Device type, e.g. B260A-13-F (SMA package); B240-13-F (SMB package).

Notes: 1. EU Directive 2002/95/EC (RoHS). All applicable RoHS exemptions applied, see EU Directive 2002/95/EC Annex Notes.
2. Product manufactured with Data Code 0924 (week 24, 2009) and newer are built with Green Molding Compound.
3. For packaging details, go to our website at http://www.diodes.com.

Marking Information

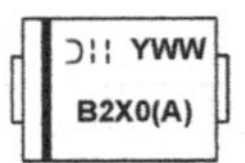

B2X0A = Product type marking code, ex: B220A (SMA package)
B2X0 = Product type marking code, ex: B230 (SMB package)
= Manufacturers' code marking
YWW = Date code marking
Y = Last digit of year (ex: 2 for 2002)
WW = Week code (01 to 53)

B220/A - B260/A
Document number: DS13004 Rev. 16 - 2

1 of 4
www.diodes.com

September 2010
© Diodes Incorporated

Figure 5.4. *Extract from diode documentation (Diodes Inc.)*

This enables us to calculate the capacitance:

$$C = \frac{\alpha.I_s}{\Delta v_s.F_d} = \frac{0.5}{0.3 \times 50000} \simeq 33\,\mu\text{F} \qquad [5.5]$$

A normalized value for the capacitor of around $68\,\mu$F can therefore be used. We then need to identify capacitors with a nominal operating voltage of over 15 V. The normalized value closest to 30 V is 35 V (30 V capacitors are also available, but there are fewer possible choices). One possible solution is shown in Figure 5.5 with aluminum electrolyte capacitors in SMD format, produced by Vishay.

However, the output ripple must be shown to conform to requirements, in spite of imperfections. Capacitors do not behave in the same way as ideal (pure) capacitances, but include an equivalent series resistance (ESR), linked to $\tan\delta$, according to the Vishay documentation (see Figure 5.6), by the relationship:

$$\text{ESR} = \frac{\tan\delta}{2\pi F_d C} \qquad [5.6]$$

In the case of the $100\,\mu$F/35 V capacitor, we have an ESR of 15.7 mΩ at 100 Hz. Note that this very low value appears to correspond to the aforementioned requirements, but it is important to remember that this value is liable to change at different frequencies. Verifications may be carried out by considering the indication of the capacitor impedance at 100 kHz, with a value of 0.4 Ω. We may consider that, at this frequency, only the resistive part of the impedance remains (unless an inductive behavior emerges); in this case, the ESR is 0.4 Ω. We therefore have an increase in ESR by a factor of between 30 and 40 across three decades of frequency range. At switching period level, a resistance-based voltage ripple of the form $ESR.I_{AC}$ may be expected with:

$$I_{AC} = \sqrt{I_{\text{DRMS}}^2 - I_s^2} = 1.29\,\text{A} \qquad [5.7]$$

140 CRH

Vishay BCcomponents

Aluminum Capacitors
SMD (Chip), High Temperature

Fig.1 Component outline

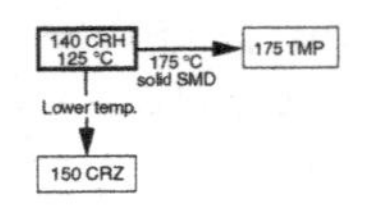

QUICK REFERENCE DATA	
DESCRIPTION	VALUE
Nominal case sizes (L x W x H in mm)	8 x 8 x 10 to 18 x 18 x 21
Rated capacitance range, C_R	10 µF to 4700 µF
Tolerance on C_R	± 20 %
Rated voltage range, U_R	6.3 V to 63 V
Category temperature range	- 55 °C to + 125 °C
Endurance test at 125 °C	1000 h to 3000 h
Useful life at 125 °C	1500 h to 5000 h
Useful life at 40 °C 1.8 x I_R applied:	150 000 h to 350 000 h
Shelf life at 0 V, 125 °C	1000 h
Based on sectional specification	IEC 60384-18/CECC 32300
Climatic category IEC 60068	55/125/56

FEATURES

- Polarized aluminum electrolytic capacitors, non-solid electrolyte, self healing
- SMD-version with base plate, lead (Pb)-free reflow solderable
- Charge and discharge proof, no peak current limitation
- Parts for advanced high temperature reflow soldering according to JEDEC J-STD-020 available
- Compliant to RoHS directive 2002/95/EC
- Vibration proof, 4-pin version and 6-pin version
- AEC-Q200 qualified

Pb-free RoHS COMPLIANT

APPLICATIONS

- SMD technology, for high temperature reflow soldering
- Industrial and professional applications
- Automotive, general industrial, telecom
- Smoothing, filtering, buffering

MARKING

- Rated capacitance (in µF)
- Rated voltage (in V)
- Date code, in accordance with IEC 60062
- Black mark or '-' sign indicating the cathode (the anode is identified by bevelled edges)
- Code indicating group number (H)

PACKAGING

Supplied in blister tape on reel

www.vishay.com 66 For technical questions, contact: aluminumcaps1@vishay.com Document Number: 28396 Revision: 23-Apr-10

Figure 5.5. *Extract from capacitor documentation (Vishay)*

In order to ensure that the ripple remains well below that generated by the capacitance alone, $ESR.I_{AC}$ must be maintained at a level considerably lower than 0.3 V (for example 0.03 V). From this, we deduce that the target ESR is of the order of 23 mΩ. However, the greater the reduction in ESR, the higher the capacitance, and the purely capacitive-based ripple is therefore reduced. The combination of these two effects make dimensioning more complicated. Nevertheless, it is possible to verify that a capacitor of 330 μF will satisfy the specification, with:

– a "capacitive" voltage ripple of 30.3 mV;

– a "resistive" voltage ripple of 0.155 V (RMS).

VISHAY. **140 CRH**

Aluminum Capacitors
SMD (Chip), High Temperature

Vishay BCcomponents

ELECTRICAL DATA AND ORDERING INFORMATION

U_R (V)	C_R (µF)	NOMINAL CASE SIZE L x W x H (mm)	I_R 100 kHz 125 °C (mA)	I_{L2} 2 min (µA)	tan δ 100 Hz	Z 100 kHz 20 °C (Ω)	ORDERING CODE [(1)] MAL2140...	ORDERING CODE [(2)] MAL2140...
35	68	8 x 8 x 10	180	24	0.14	0.40	97003E3	-
	100	10 x 10 x 10	255	35	0.14	0.25	97001E3	99001E3
	150	10 x 10 x 14	317	53	0.14	0.20	97002E3	-
	220	12.5 x 12.5 x 13	750	77	0.14	0.12	97011E3	99011E3
	330	12.5 x 12.5 x 13	750	115	0.14	0.12	97012E3	99012E3
	470	12.5 x 12.5 x 16	900	164	0.14	0.09	97013E3	99013E3
	680	16 x 16 x 16	1100	238	0.14	0.08	-	99014E3
	820	16 x 16 x 21	1200	287	0.14	0.06	-	99015E3
	1000	18 x 18 x 16	1200	350	0.14	0.08	-	99016E3
	1200	18 x 18 x 21	1550	420	0.14	0.06	-	99017E3
	1500	18 x 18 x 21	1550	525	0.14	0.06	-	99018E3
50	47	8 x 8 x 10	145	24	0.14	0.70	97103E3	-
	68	10 x 10 x 10	205	34	0.14	0.50	97101E3	99101E3
	100	10 x 10 x 14	255	50	0.14	0.40	97102E3	-
	220	12.5 x 12.5 x 13	750	110	0.12	0.23	97111E3	99111E3
	330	12.5 x 12.5 x 16	900	165	0.12	0.18	97112E3	99112E3
	470	16 x 16 x 16	900	235	0.12	0.15	-	99113E3
	680	16 x 16 x 21	1000	340	0.12	0.13	-	99114E3
	680	18 x 18 x 16	1000	340	0.12	0.15	-	99115E3
	820	18 x 18 x 16	1000	410	0.12	0.15	-	99116E3
	820	18 x 18 x 21	1050	410	0.12	0.13	-	99117E3
	1000	18 x 18 x 21	1050	500	0.12	0.13	-	99118E3
63	10	8 x 8 x 10	145	6.3	0.12	0.70	97805E3	-
	22	8 x 8 x 10	145	14	0.12	0.70	97803E3	-
	33	8 x 8 x 10	145	21	0.12	0.70	97804E3	-
	47	10 x 10 x 10	205	30	0.12	0.50	97801E3	99801E3
	68	10 x 10 x 14	255	43	0.12	0.40	97802E3	-
	100	12.5 x 12.5 x 13	500	63	0.10	0.25	97811E3	99811E3
	220	12.5 x 12.5 x 16	600	138	0.10	0.20	97812E3	99812E3
	330	16 x 16 x 16	700	208	0.10	0.18	-	99813E3
	330	16 x 16 x 21	750	208	0.10	0.15	-	99814E3
	330	18 x 18 x 16	750	208	0.10	0.18	-	99815E3
	470	16 x 16 x 21	750	296	0.10	0.15	-	99816E3
	470	18 x 18 x 16	750	296	0.10	0.18	-	99817E3
	470	18 x 18 x 21	900	296	0.10	0.15	-	99818E3
	680	18 x 18 x 21	900	428	0.10	0.15	-	99819E3

Notes

(1) Standard reflow soldering profile, see fig.4 and table 4

(2) Advanced reflow soldering profile, according to JEDEC J-STD-020, see fig.5 and table 5

Document Number: 28396
Revision: 23-Apr-10

For technical questions, contact: aluminumcaps1@vishay.com

www.vishay.com
73

Figure 5.6. *Extract from capacitor documentation (Vishay) (contd.)*

This gives a total of 0.230 V (maximum), which conforms to the requirements set out in the specification. It is best to maintain a safety margin, as the resistive ripple is calculated based on an RMS value of the alternating current in the capacitor and not on the "peak" value, which is equal to 3 A (1 A of average current in the load and 4 A peak). The instantaneous voltage peak induced by a step of 3 A in the ESR of the resistance gives us a voltage of 0.36 V: this voltage is too high for the specifications.

SUMMARY 5.1.– Finally, we note that the decoupling capacitor needs to be dimensioned on the basis of the ESR. The 330 μF capacitor does not offer a sufficiently comfortable safety margin, so a 470 μF will be preferred in this case; this new component is not much larger (in terms of height) and the issue is not critical if we use a transformer based on a classic ferrite core (and not a "planar" type core). This transformer will be studied in the following section. The sizing process has demonstrated that the "capacitive" voltage ripple is not particularly important when dimensioning an electrolyte capacitor; it is better to calculate the ripple resulting from the ESR. Moreover, the decoupling may be considerably improved (with a consequent reduction in the failure risk) by associating the electrolyte capacitor with a non-polarized capacitance (film or ceramic), placed in series; this capacitance plays a role in HF mode, and may be dimensioned so as to maintain a low impedance over a range of high frequencies. This result may be obtained using the Taiyo Yuden 10 μF/35 V capacitor GMK325BJ106KN-T, with a reduced volume and in SMD 1210 format (2.5 mm x 3.2 mm x 1.9 mm).

5.3.2. *Coupled inductances*

The turns ratio $m = 3$ has been defined arbitrarily. The primary inductance required for the application may be calculated based on the hypothesis of full demagnetization, as we know the switching frequency (F_d = 50 kHz) and the required output power of the converter (P_s = 15 W). The energy to store in each switching period can then be deduced:

$$W_{\mathrm{mag}} = \frac{P_s}{F_d} = 0.3\,\mathrm{mJ} \qquad [5.8]$$

Given that the maximum current in the primary (at the end of the first switching phase $\alpha.T_d = T_d/2$) is equal to 12 A ($I_{T\max}$), the required inductance L_1 can easily be calculated, as:

$$W_{\mathrm{mag}} = \frac{1}{2} L_1.I_{T\max}^2 \qquad [5.9]$$

hence:

$$L_1 = \frac{2W_{\mathrm{mag}}}{I_{T\max}^2} \simeq 4,2\,\mu\mathrm{H} \qquad [5.10]$$

Given that $L_1 = a_L.n_1^2$ and $n_2/n_1 = 3$, we can also state that $L_2 = 37.5\,\mu\mathrm{H}$ if the magnetic coupling between the primary and the secondary is perfect.

5.4. Dimensioning coupled inductances

The input data for the coupled inductance dimensioning process is obtained partly from the application specification, but also from "trade" information concerning the use of the magnetic material (B_{max}) and copper (J_{max}) and the quality of the winding (K_b). In this case, we will use the following values:

- $B_{\max} = 0.2\,\mathrm{T}$;
- $J_{\max} = 5\,\mathrm{A/mm^2} = 5 \times 10^6\,\mathrm{A/m^2}$;
- $K_b = 0.5$.

Given the limitations of the specification with an output power $\mathcal{P}$ of 15 W and a switching frequency F_d of 50 kHz, we can carry out the numerical application corresponding to the minimum required product $A_e.S_b$:

$$A_e.S_b \geq 2\sqrt{\frac{2}{3}}\frac{\mathcal{P}}{B_{\max}.K_b.J_{\max}.F_d} = 9.8 \times 10^{-10}\,\mathrm{m}^4 = 98\,\mathrm{mm}^4 \qquad [5.11]$$

5.4.1. *Choice of a ferrite core*

The first stage in selecting a ferrite core is to identify the appropriate family of cores. In power electronics, the most interesting cores "surround" the winding, insofar as they prevent magnetic leakage from the component; this leakage can cause parasite effects for the closest elements on a PCB. There are two distinct families in this category of cores, as shown in Figure 5.7: RM cores (left) and PM cores (right).

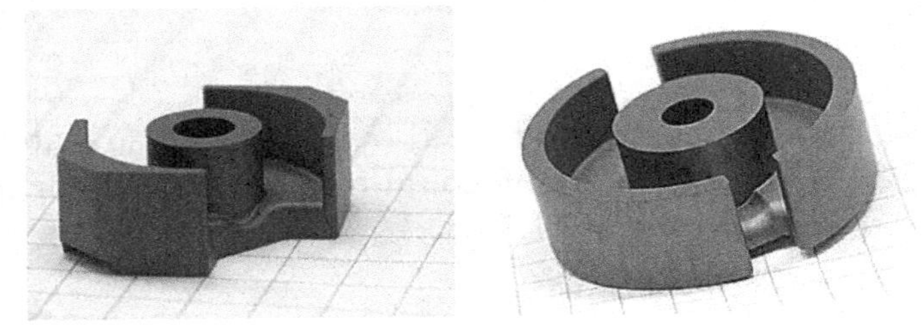

Figure 5.7. *Low-radiation ferrite cores (left: RM; right: PM – source: Kolektor Magma)*

In this case, we have arbitrarily chosen an RM type core. We now need to determine the size required for our application. This is carried out using the values of A_e, S_b and the product $A_e.S_b$ of the full RM range (from the smallest, RM4, to the largest, RM14/ILP) as shown in Table 5.2.

Based on these results, any of the cores may be used; however, a certain space margin should be left for the winding, which will have an impact both on the coupling between the primary and secondary windings and on the optimization (or otherwise) of the windable window. In our winding coefficient, we have taken account of the fact that empty spaces always exist between conductors (with a circular cross-section) and due to the plastic frame (generally made of nylon) used for windings, which are never created directly on the ferrite core itself. For this reason, we should, *a priori*, consider a core of type RM4, RM4/I, RM5, RM5/I or RM5/ILP. While the latter options are considerably overdimensioned, it is important to note that the footprint of a converter does not simply concern the volume (in cm^3), but also the location. Depending on the application, an over-dimensioned core may be chosen due to its low profile (LP) type geometry.

<table>
<tr><th>Size</th><th>A_e (in mm^2)</th><th>S_b (in mm^2)</th><th>$A_e.S_b$ (in mm^4)</th></tr>
<tr><td>RM4</td><td>11</td><td rowspan="2">14.2</td><td>156.2</td></tr>
<tr><td>RM4/I</td><td>13.8</td><td>267</td></tr>
<tr><td>RM4/ILP</td><td>14.5</td><td>8.71</td><td>126.3</td></tr>
<tr><td>RM5</td><td>21.2</td><td rowspan="2">16.7</td><td>354</td></tr>
<tr><td>RM5/I</td><td>24.8</td><td>414.2</td></tr>
<tr><td>RM5/ILP</td><td>24.5</td><td>9.54</td><td>233.7</td></tr>
<tr><td>RM6S</td><td>31.4</td><td rowspan="2">24</td><td>753.6</td></tr>
<tr><td>RM6S/I</td><td>37</td><td>888</td></tr>
<tr><td>RM6S/ILP</td><td>37.5</td><td>13.5</td><td>506.3</td></tr>
<tr><td>RM6R</td><td>32</td><td>24</td><td>768</td></tr>
<tr><td>RM7/I</td><td>44.1</td><td>31.5</td><td>1389</td></tr>
<tr><td>RM7/ILP</td><td>45.3</td><td>19.2</td><td>869.8</td></tr>
<tr><td>RM8</td><td>52</td><td rowspan="2">45.6</td><td>2371</td></tr>
<tr><td>RM8/I</td><td>63</td><td>2873</td></tr>
<tr><td>RM8/ILP</td><td>64.9</td><td>24.9</td><td>1616</td></tr>
<tr><td>RM10/I</td><td>96.6</td><td>63.9</td><td>6173</td></tr>
<tr><td>RM10/ILP</td><td>99.1</td><td>34.5</td><td>3419</td></tr>
<tr><td>RM12/I</td><td>146</td><td>102.5</td><td>14970</td></tr>
<tr><td>RM12/ILP</td><td>148</td><td>54.9</td><td>8125</td></tr>
<tr><td>RM14/I</td><td>198</td><td>145.6</td><td>28830</td></tr>
<tr><td>RM14/ILP</td><td>201</td><td>77.7</td><td>15620</td></tr>
</table>

Table 5.2. *Geometric characteristics of the RM family of cores (source: Ferroxcube)*

In this case, if component height is a priority, it is best to use LP-type components. However, the gain is relatively modest, and on closer inspection of the technical documentation, these cores are not available with an air gap (required in a Flyback power supply, which is based on the storage of magnetic energy). For this reason, an RMx or RMx/I type core is required. The difference between these two cores lies in their central axis, which may (RMx core) or may not (RMx/I) include a hole. The two cores are geometrically identical: however, the RMx cores are lighter, while still offering a significant margin in relation to the product $A_e.S_b$, given our minimum required value of $98\,\text{mm}^4$. An extract from the Ferroxcube documentation for an RM5 core is provided for illustrative purposes in Figure 5.8, with different specific inductance values a_L as a function of the air gap.

Once a core has been selected, the following choices must be shown to conform to the specification, particularly in terms of maximum magnetic flux density. We know that the application requires an inductance L_1 of $4.2\,\mu\text{H}$. Using relationship $L_1 = a_L.n_1^2$, it is possible to calculate the number of turns required for all available specific inductances:

$$n_1 = \sqrt{\frac{L_1}{a_L}} \tag{5.12}$$

A summary of the results is shown in Table 5.3 with an indication of the corresponding maximum magnetic flux density, given that we have:

$$n_1.I_{1\max} = \mathcal{R}.B_{\max}.A_e = \frac{B_{\max}.A_e}{a_L} \tag{5.13}$$

hence:

$$B_{\max} = \frac{n_1.a_L.I_{1\max}}{A_e} \tag{5.14}$$

and:

– $I_{1\max} = 12\,\mathrm{A}$ for this converter;

– $A_e = 21.2\,\mathrm{mm}^2$ with the RM5 core.

REMARK 5.1.– With the exception of the case where the calculated number of turns is of the form $x.0\cdots$ (value rounded to x), the values shown in Table 5.3 are rounded to $x+1$.

Ferroxcube

RM, RM/I, RM/ILP cores and accessories — RM5

CORE SETS

Effective core parameters

SYMBOL	PARAMETER	VALUE	UNIT
$\Sigma(l/A)$	core factor (C1)	1.01	mm^{-1}
V_e	effective volume	450	mm^3
l_e	effective length	21.4	mm
A_e	effective area	21.2	mm^2
A_{min}	minimum area	14.8	mm^2
m	mass of set	≈ 3.1	g

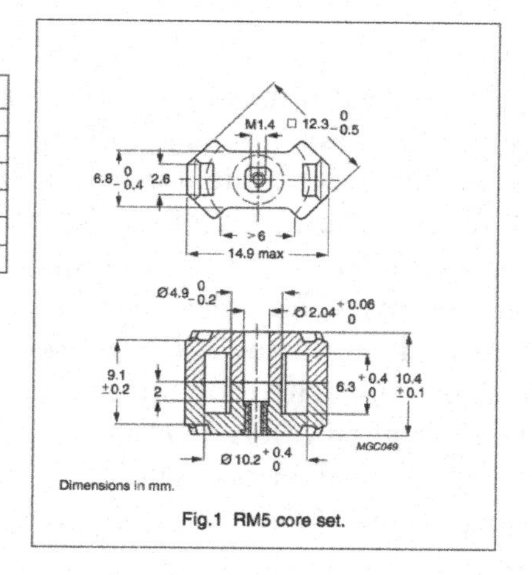

Dimensions in mm.

Fig.1 RM5 core set.

Core sets for filter applications

Clamping force for A_L measurements, 25 ±10 N.

GRADE	A_L (nH)	μ_e	TOTAL AIR GAP (μm)	TYPE NUMBER (WITH NUT)	TYPE NUMBER (WITHOUT NUT)
3D3 sup	40 ±3%	≈32	≈990	RM5-3D3-E40/N	RM5-3D3-E40
	63 ±3%	≈51	≈540	RM5-3D3-E63/N	RM5-3D3-E63
	100 ±3%	≈80	≈300	RM5-3D3-E100/N	RM5-3D3-E100
	800 ±25%	≈640	≈0	–	RM5-3D3
3H3 sup	160 ±3%	≈129	≈180	RM5-3H3-A160/N	RM5-3H3-A160
	250 ±3%	≈201	≈110	RM5-3H3-A250/N	RM5-3H3-A250
	315 ±3%	≈253	≈80	RM5-3H3-A315/N	RM5-3H3-A315
	400 ±5%	≈321	≈60	RM5-3H3-A400/N	RM5-3H3-A400
	1650 ±25%	≈1310	≈0	–	RM5-3H3

Figure 5.8. *Extract from RM5 core documentation (source: Ferroxcube)*

Material-Grade / Air gap (in μm)	a_L (in nH)	n_1 required	$B_{\max}$ (in T)
3D3/990μm	40	11	0.249
3D3/540μm	63	9	0.321
3D3/300μm	100	7	0.393
3D3/0μm	800	3	1.36
3H3/180μm	160	6	0.543
3H3/110μm	250	4	0.566
3H3/80μm	315	4	0.713
3H3/60μm	400	4	0.906
3H3/0μm	1,650	2	1.87

Table 5.3. *Number of turns n_1 required as a function of the specific inductance a_L of the core and the corresponding magnetic flux density $B_{\max}$ (@ $i_1 = I_{1\max} = 12$ A)*

The maximum magnetic flux densities are too high. A magnetic flux density induction slightly higher than $0.2\,\mathrm{T}$ might be acceptable, but even the lowest air gap value (grade 3D3) is close to $0.25\,\mathrm{T}$, which is the "saturation knee" of the material at 100°C. This operating mode would generate nonlinearity, but also lead to excessive losses, with a risk of thermal runaway. While the RM5 core is theoretically compatible with our specifications and physical constraints, the air gaps offered by the manufacturer do not allow us to obtain the desired result. We therefore need to consider a larger core. Generally speaking, the presence of the air gap (with normalized available values) means that the dimensioning of a Flyback transformer is an iterative process, and is more "complicated" than for the Forward transformer.

The calculation approach may easily be inverted in order to verify whether the specific inductances offered for an RMx core model are suitable. To do this, we use the equation:

$$n_1 = \frac{L_1.I_{1\max}}{B_{\max}.A_e} \qquad [5.15]$$

The number of turns required can be established for all cores compatible with the constraint regarding the product $A_e.S_b$ (see Table 5.4).

Core	A_e(in mm^2)	n_1 required	a_L required (in nH)
RM6S	31.4	8	65.6
RM6S/I	37	7	85.7
RM6S/ILP	37.5	7	85.7
RM6R	32	8	65.6
RM7/I	44.1	6	117
RM7/ILP	45.3	6	117
RM8	52	5	168
RM8/I	63	4	263
RM8/ILP	64.9	4	263

Table 5.4. *Number of turns n_1 required as a function of the specific inductance a_L of the core and the corresponding magnetic flux density $B_{\max}$ (@ $i_1 = I_{1\max} = 12$ A)*

Clearly, we require a considerably larger core to that deduced based on the product $A_e.S_b$ alone, in order to guarantee satisfactory levels of maximum magnetic flux density in the material. An RM8 core (a version with a specific inductance of 160 nH is available) or an RM8/ILP core (with $a_L = 250$ nH) may therefore be used. The second solution is preferable, as the profile of the core is lower, making it interesting when designing a compact power supply. The reduction in the winding cross-section is not problematic, as the core is considerably over-dimensioned from this perspective. An extract from the documentation for this component is shown in Figure 5.9.

5.4.2. *Windings*

The summary of the RM8/ILP core selection process in the previous section provided us with a full characterization of the magnetic circuit, with the choice of an air gap suited to the specification and to the physical limitations of the material. For our study, we have selected model RM8/ILP-3D3-A250, with a specific inductance a_L of 250 nH. Thus, if we wish to obtain an inductance L_1 of $4,2\mu$H, the number of turns required on the primary winding is $n_1 = 4$. In practice, using this rounded value (slightly lower than the required value), we obtain an inductance of $4\,\mu$H (around 5% lower than the theoretical value). This difference is not a problem, as it is always possible to adjust the switching frequency in order to enter critical mode, and compensate for the slight reduction in stored energy during each switching period. However, this choice does guarantee that the material will not become saturated, facilitating the converter to operate (*a priori*) with no risk of thermal runaway. Once the number of turns on the primary conductor has been selected, the value of n_2 is also defined as, from the outset, we have stated that $n_2/n_1 = 3$ (hence $n_2 = 12$).

We still need to define the required cross-section of the conductors and choose an appropriate standardized cross-section: *a posteriori* verification of the fullness of the windable window S_b will therefore be required.

The RMS values of currents are used to calculate conductor cross-sections:

– $I_{1\text{RMS}} = I_{T\text{RMS}} = 4.90\,\text{A}$;

– $I_{2\text{RMS}} = I_{D\text{RMS}} = 1.63\,\text{A}$.

With an acceptable current density of $5\,\text{A/mm}^2$ in the conductors, we require cross-sections of:

– $0.98\,\text{mm}^2$ in the primary winding;

– $0.326\,\text{mm}^2$ in the secondary winding.

Ferroxcube

RM, RM/I, RM/ILP cores and accessories RM8/ILP

CORE SETS

Effective core parameters

SYMBOL	PARAMETER	VALUE	UNIT
Σ(l/A)	core factor (C1)	0.440	mm⁻¹
V_e	effective volume	1860	mm³
l_e	effective length	28.7	mm
A_e	effective area	64.9	mm²
A_{min}	minimum area	55.4	mm²
m	mass of set	≈ 10	g

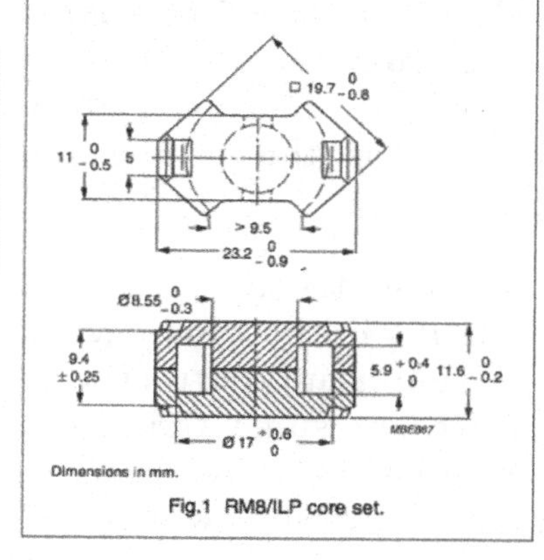

Fig.1 RM8/ILP core set.

Core sets for filter applications

Clamping force for A_L measurements, 30 ±10 N.

GRADE	A_L (nH)	μ_e	AIR GAP (μm)	TYPE NUMBER
3B46 des	6500 ±25%	≈ 2280	≈ 0	RM8/ILP-3B46
3D3	250 ±3%	≈ 88	≈ 330	RM8/ILP-3D3-A250
	315 ±3%	≈ 111	≈ 250	RM8/ILP-3D3-A315
	400 ±5%	≈ 141	≈ 180	RM8/ILP-3D3-A400
	1850 ±25%	≈ 650	≈ 0	RM8/ILP-3D3
3H3	400 ±3%	≈ 141	≈ 210	RM8/ILP-3H3-A400
	630 ±5%	≈ 222	≈ 120	RM8/ILP-3H3-A630
	1000 ±8%	≈ 352	≈ 70	RM8/ILP-3H3-A1000
	4100 ±25%	≈ 1440	≈ 0	RM8/ILP-3H3

Core sets for general purpose transformers and power applications

Clamping force for A_L measurements, 30 ±10 N.

GRADE	A_L (nH)	μ_e	AIR GAP (μm)	TYPE NUMBER
3C90	4100 ±25%	≈ 1440	≈ 0	RM8/ILP-3C90
3C94	4100 ±25%	≈ 1440	≈ 0	RM8/ILP-3C94
3C95 des	4800 ±25%	≈ 1690	≈ 0	RM8/ILP-3C95
3C96 des	3800 ±25%	≈ 1330	≈ 0	RM8/ILP-3C96
3F3	3800 ±25%	≈ 1330	≈ 0	RM8/ILP-3F3
3F35 prot	3100 ±25%	≈ 1090	≈ 0	RM8/ILP-3F35
3F4 des	2200 ±25%	≈ 770	≈ 0	RM8/ILP-3F4
3F45 prot	2200 ±25%	≈ 770	≈ 0	RM8/ILP-3F45

2008 Sep 01 2

Figure 5.9. *Extract from the documentation for the selected core (RM8 / ILP – source: Ferroxcube)*

Note that the skin thickness in copper at 50 kHz may be calculated using a formula already seen above:

$$\delta_f = \sqrt{\frac{2}{\omega\mu\sigma}} \qquad [5.16]$$

where $\mu = \mu_0 = 4\pi \times 10^{-7}\,\text{H/m}$, $\sigma = \sigma_{Cu} = 59,6 \times 10^6\,\text{S/m}$ and $\omega = 2\pi F_d = 2\pi \times 5 \times 10^4\,\text{rad/s}$. This results in a skin thickness δ_f of $0.29\,\text{mm}$. The maximum cross-section of a wire can then be calculated based on copper with a diameter of $2\delta_f$. This gives a cross-section of $0.267\,\text{mm}^2$. Conductors therefore need to be split in order to improve efficiency, as the cross-sections required for the primary and secondary conductors are higher.

It is not strictly correct to consider that all of the current circulating in the windings will be affected by the skin effect, as a significant part of the current is constant. We will therefore presume that the secondary conductor does not need to be split, and a cross-section directly superior to the value of $0.326\,\text{mm}^2$ will be used (in this case, a wire with a cross-section of $0.4\,\text{mm}^2$, reference ECW0.71, produced by PRO POWER). However, to ensure correct operation of the primary winding, three twisted parallel strands of this wire are used[3]. This multi-strand winding can be produced in a satisfactory manner by clamping one end of the three strands in a vise, and placing the other end of the strands in a drill chuck. By placing sufficient mechanical tension on the three wires (which should all be correctly tensed) and rotating the drill, a regular twist may be obtained, as shown in Figure 5.10.

5.4.3. *Tests and leakage measurements*

Once the parameters of the transformer have been fully determined (choice of a core – including the air gap, calculation of the numbers of turns n_1 and n_2, identification of the required cross-section of conductors and wire splitting, where necessary), the next stage in the converter design process is to create the windings for the coupled inductances,

3 This solution is an effective response to the skin effect, and much more affordable than the use of Litz wire, as mentioned in Chapter 2.

and to identify one key parameter, which cannot be determined by theoretical means: the leakage inductance of the transformer. The quality of the coupling between the two windings is dependent on a number of factors:

– the quality of the winding;

– the ease of production of the winding, based on the required cross-sections (which are often very different);

– the number of turns required;

– the way in which the two windings are associated on the base.

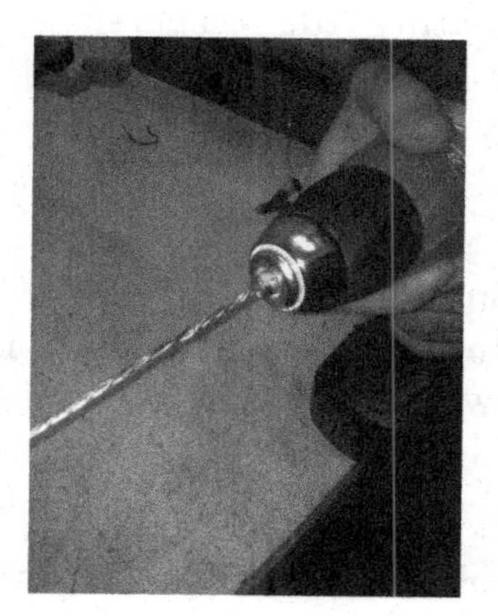

Figure 5.10. *Economical production of a conductor with three isolated twisted strands*

A highly detailed study of these aspects of transformer design is given in [FER 02]. In this case, we will simply present the leakage inductance identification process. To do this, a power source (such as a GBF) is needed; it should be possible to set the frequency of the source to a value close to the switching frequency used for the converter (in this case, 50 kHz). For the purpose of this study, we will consider operations in sinusoidal mode, which is the classic configuration for measurements in a transformer (this is equivalent to the study of a 50 Hz transformer). In this case, the magnetizing inductance is low, and it is *a priori* not possible to identify a leakage inductance through short-circuit testing. However, we can easily carry out open circuit identification of the inductance L_1, powering the primary with the secondary left open ($i_2 = 0$) and observing the resulting voltage v_2. This enables identification of the turns ratio m, and consequently the mutual inductance M between the two windings. A symmetrical process for the secondary conductor (with the primary left open) can then be carried out to identify L_2 (this test also allows re-evaluation of the mutual inductance M). Once these parameters have been established[4], we may evaluate the quality of the coupling by calculating the dispersion coefficient of the transformer:

$$\sigma = 1 - \frac{M^2}{L_1 L_2} \qquad [5.17]$$

In practice, a coupling with a dispersion of the order of 5% may be expected. In this case, the leakage inductance totaled in the primary is written σL_1; in our case, it has a value of $0.05 \times 4\mu\text{H} = 200\,\text{nH}$.

5.5. Transistor control and snubber calculation

5.5.1. *Determining gate resistance*

To determine the gate resistance needed for switching in the transistor, we begin by analyzing the expected operation of the converter. We know that the converter must operate at a switching frequency of 50 kHz, i.e. with a switching period of 20 μs. A transistor with a switching period of 200 ns is therefore selected, so that these switching operations will be

4 In practice, it is also possible to evaluate the resistances of the windings, which should be low if conductors have been correctly split in order to minimize the skin effect.

negligible in relation to the switching period. Once the switching time has been defined ($t_{on} = t_{off} = 200\,\text{ns}$), a suitable gate resistance is determined as a function of the gate capacity and the power voltage. In this case, we will assume that the grid is controlled using 5 V. We then use characteristic $V_{GS} = f(Q_G)$ of the transistor, provided by the manufacturer (see Figure 5.11). Using 5 V indicates that the charge stored by the transistor gate is of the order of 10 nC. Given that the capacitance C_G of the gate may be considered to be equal to 2 nF (this is, in fact, a variable capacitance) using the relationship $Q_G = C_G.V_{GS}$, a suitable gate resistance may be evaluated by considering, as a first approximation, that $t_{on} = t_{off} = 3\tau = 3R_GC_G$. A gate resistance of around 33 Ω is therefore required. At this point, we note that the gate control does not require a significant pulsed current (approximately 150 mA maximum): therefore it should be easy to find a driver suited to this function.

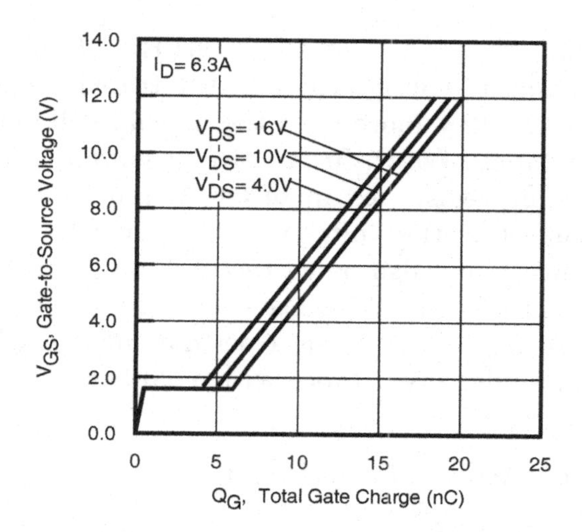

Figure 5.11. *$V_{GS} = f(Q_G)$ characteristic of the transistor (source: ST Microelectronics)*

REMARK 5.2.– Note, however, that the calculation of the gate resistance should be refined using a SPICE-type simulation, in order to give sufficient consideration to the real (nonlinear) behavior of the transistor.

5.5.2. *RCD snubber circuit*

It is then easy to calculate the magnetic energy W_f stored in the primary leakage inductance of the transformer at the moment of switching, as we know that current i_1 has a value of 12 A at this point:

$$W_f = 1/2.\sigma L_1.I_{1\max}^2 = 14.4\mu\mathrm{J} \qquad [5.18]$$

We have shown that for an RCD snubber, we need to include a resistance R with a value expressed as:

$$R = \frac{(V_{\max} - V_e)^2}{W_f.F_d} \qquad [5.19]$$

Given that $V_e = 5\,\mathrm{V}$ and $F_d = 50\,\mathrm{kHz}$, we must simply choose a maximum voltage value at the transistor terminals, for example $12\,\mathrm{V}$ ($2\,\mathrm{V}$ more than the normal voltage $V_e + \frac{V_s}{m}$), given a resistance of $58\,\Omega$. In practice, it is better to choose a slightly weaker resistance in order to reduce the snubber output voltage (i.e. the voltage at the capacitor terminals); this resistance must also be correctly dimensioned in power terms. In this case, the resistance should be able to withstand a power of $W_f.F_d = 0.72\,\mathrm{W}$ on a permanent basis; we might select, for example, a resistance with a caliber of $1\,\mathrm{W}$.

5.6. PWM control and regulation

5.6.1. *PWM controller*

In this case, PWM control is carried out by a dsPIC33F06GS101 type microcontroller (Microchip). This family of components is designed for the control of switch mode power supplies, with clocking carried out by a quartz element at 10 MHz and a PLL, internally multiplying this frequency by 8. This allows treatment of 37.5 MIPS (MIPS = million instructions per second). The peripherals in the microcontroller include timers, which may be used to generate one or more PWM outputs. In this case, this function is used to control the single transistor in the Flyback structure, and we have fixed the switching frequency at 50 kHz (a frequency outside the audible spectrum, but sufficiently low to avoid interactions with parasite components linked to non-optimal card routing; this frequency also limits switching losses and iron losses). This means that the device will generate a lag in the model of the regulation loop; we will suppose (for safety reasons) that this lag is equal to a full switching period ($e^{-T_d s}$) rather than to the lag $e^{-\frac{T_d s}{2}}$ introduced in Chapter 4. An alternative to the microcontroller would be the use of a dedicated Pulse Width Modulator Control circuit, such as the SG3525 available from ON Semiconductor.

REMARK 5.3.– A pure delay of 200 μs is negligible in practice, when considered in relation to the other time constants in the system (particularly the converter output filter). It can therefore be ignored when parameterizing the corrector used for closed-loop operations.

5.6.2. *Galvanic isolation of controls*

The measurement of the converter output voltage must be transmitted to the control circuit, entirely situated on the "primary" side. We have developed a power structure including galvanic isolation, and this property needs to be conserved at control level: the output voltage measurement should not, therefore, be directly transmitted through a wire to the microcontroller on the primary side. The chosen solution (see Figure 5.12) is based on an integrated IL300 linear optocoupler, developed by Vishay.

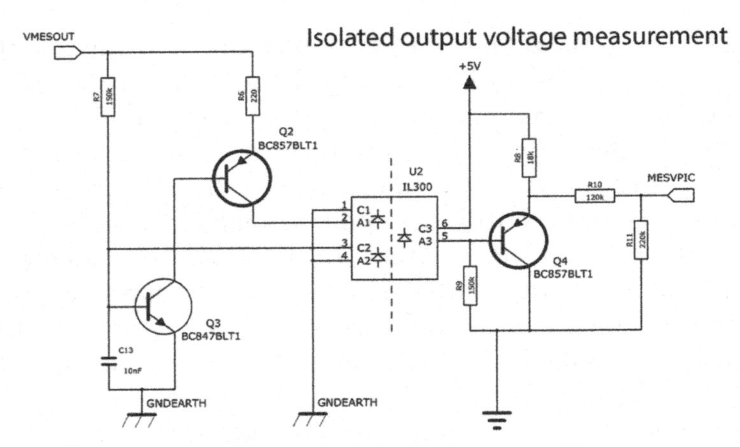

Figure 5.12. *Block diagram of the linear optocoupler circuit*

Despite the “linear” element of the name, this optocoupler is based on an emitting diode and on photodiodes (used as receivers) and is thus fundamentally nonlinear. The linearity of the global characteristic of the optocoupler is based on two key elements:

– the closed-loop operation of the device;

– the coupling between the emitting diode and the identical characteristics of the photodiodes.

On this basis, the operation of the optocoupler may be analyzed using a block diagram. We have seen that the emitter is controlled based on a signal captured by one of the receiving diodes. A current quasi-proportional to the VMESOUT voltage (VMESOUT/R7 if we ignore the VBE voltage of the transistor) is injected into the base of transistor Q3. From this current, we subtract the current circulating in the receiving diode, with terminals denoted as C2 (3) and A2 (4). The difference between these currents is then multiplied by the product of the gains of transistors Q3 and Q2 (presumed to be identical in this case, and denoted as β) to obtain the current circulating in the emitting diode. This structure means that the diagram clearly represents the closed-loop structure shown in Figure 5.13. The second photodiode is placed directly in line, and is linked to the same current, denoted as i_{coll}. The nonlinear blocks represented on the block diagram describe the emitting/receiving diode assembly (DE/R), which we will consider to be strictly identical in both cases; manufacturers are able to produce this identity by construction and by integration in the same casing, with a temperature which is as uniform as possible.

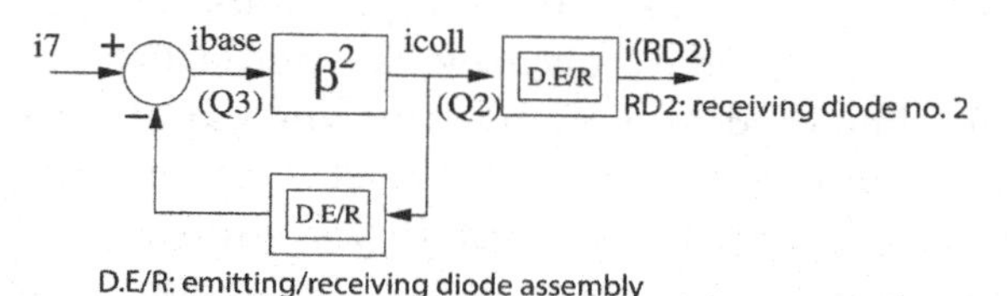

Figure 5.13. *Block diagram of the linear optocoupler circuit*

We know that a closed-loop system with a gain in the action chain behaves as the inverse function of the return path: this is true of both linear and nonlinear systems. Overall, we therefore obtain a nonlinearity in cascade with the inverse nonlinearity. Thus, the structure behaves, globally, with a unitary gain over a range of frequencies including the "continuous" element. This is our aim in this case, as we wish to regulate a voltage which should be constant. In practice, the bandwidth (upper limiting frequency) of our regulation is limited by the bandwidth of the opto-isolated voltage measurement, in this case of the order of a few hundred Hertz.

5.6.3. *Notes on modeling and control*

In this study, we have already seen that the PWM controller can be assimilated to a simple gain, and the previous section showed that the isolated measurement of the output voltage, while nonlinear, may be linearized by calibration and programmed into a microcontroller. The whole regulation loop (with the exception of the controller) can therefore be summarized as a gain. We will arbitrarily fix this gain value at 1: the value can be corrected artificially in the microcontroller. We then need to establish the transfer function of the Flyback power supply, or, more precisely, the transfer functions relating to control signal α and to the input voltage v_e (which is, in fact, a disturbance input).

The broad principles of converter modeling for control purposes were discussed in the previous chapter in the context of non-isolated structures; we noted that the results obtained for Forward and Flyback power supplies were close to those for the Buck and Buck-Boost systems, respectively. From an experimental perspective, we simply (to obtain an almost perfect model) control the switch-mode power supply in an open loop when it is charged (non-infinite load resistance), producing a control signal of the form $\alpha(t) = \alpha_0 + k.\cos(\omega t + \phi_0)$, taking $\alpha_0 = 1/2$ (the static operating point selected for our application) and choosing an oscillation amplitude k which is sufficiently small for a linear model (by transfer function) of the power supply to be valid, but sufficiently high to enable observation of ripples at angular frequency ω of the output (which, in practice, is subject to noise at frequency F_d due to switching). Clearly, in this context, frequency analysis (creation of a Bode or Black diagram) of the open loop transfer function may be applied to synthesize a corrector for closed-loop operation of the power supply.

5.6.4. *Regulator tuning*

For regulator tuning purposes, a classic approach used for design in automatics (see [COH 00]) is applied, establishing closed-loop performances in terms of:

– stability (phase and/or gain margin);

– precision (static and dynamic, with a response time).

In the context of switch mode power supplies, a reasonable response time should not be too close to the switching period $T_d = 1/F_d$. In practice, we aim to establish a response time of the order of 5 or 10 times the switching period. If we then choose to cancel the static error, an integrator must be added to the regulation loop in order to achieve this result. In these conditions (increase in the loop gain in terms of rapidity and the addition of an integral action to cancel the static error), the system is liable to be highly unstable (with a low phase/gain margin, or even full instability); in these cases, a phase offset is generally used to guarantee stability, and possibly sufficiently damped transitory behavior.

5.6.5. *Production*

This circuit was built industrially on a PCB by Eurocircuits (www.eurocircuits.com)[5]. The components (mostly SMD) were reflow soldered (using a soldering paste and a reflow oven) at the Université de Technologie de Compiègne (UTC), France. The position of the components is shown in Figure 5.14, and a photograph of the finished model is shown in Figure 5.15.

The embedded program for the microcontroller was developed using MPLAB X software, distributed free of charge by Microchip, and the XC16 compiler for components PIC24 and DSPIC, from the same source (we used the free Lite version). The presence of an integrated programming port enables students to modify the program in order to adjust closed-loop controls (for example modifying the regulator).

---

5 Eurocircuits uses a pooling approach to manufacture PCBs for prototypes or small product runs.

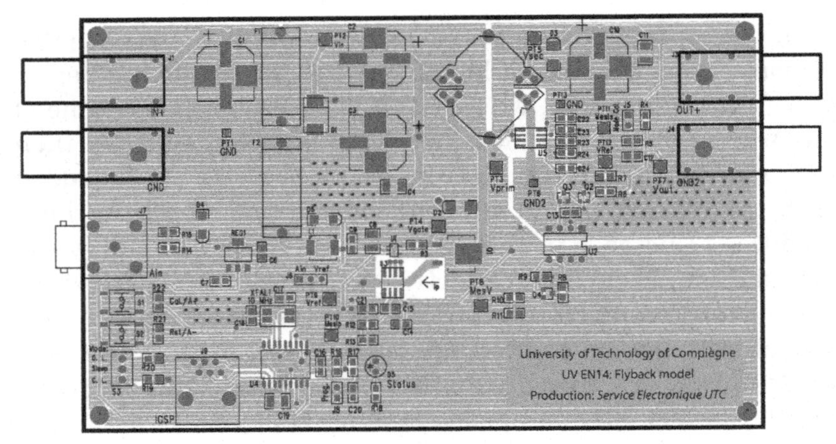

Figure 5.14. *Installation diagram for components in the Flyback power supply*

Figure 5.15. *Photograph of the finished "Flyback " evaluation board*

5.6.6. *Simulations and experimental results*

Simulation is a useful means of validating the design of a switch-mode power supply. The Simulation Program with Integrated Circuit Emphasis (SPICE) program, developed at the University of Berkeley (California) in the early 1970s, is ideal for this purpose. The software is now available in the public domain, and constitutes a *de facto* standard in the electronics industry, as most manufacturers of analog electronic components offer SPICE models of their integrated circuits (transistors, diodes, operational amplifiers, etc.). SPICE has been embedded in several industrial EDA[6] software suites (including Cadence, Mentor Graphics and Altium). These versions are distributed via software chains, which may be very costly. However, free versions are also available, such as LTSpice IV, provided by Linear Technology (the main activity of the company lies in designing integrated circuits, and especially analog circuits, as the name suggests).

Figure 5.16 shows the software user interface, with a schematic of the Flyback power supply. This illustration shows several measurement points (from top to bottom, the primary and secondary currents and the secondary voltage of the transformer). Figure 5.17 gives a more detailed view of the voltage at the transistor terminals in discontinuous conduction mode (with a resistive load of $30\,\Omega$). Compared to the theoretical basis, this illustration presents a significant difference: when the current in the diode cancels out, the voltages in the primary and secondary of the transformer should be zero, and we should therefore obtain a voltage of $5\,V$ at the transistor terminals. If the average value of this voltage is around $5\,V$, we have a high ripple, characteristic of a resonant circuit. This resonance is the result of an interaction between the transformer inductances and the parasite capacitances of the two switches in the OFF state (transistor and diode). This phenomenon is also seen in practice, and demonstrates the interest of SPICE simulation, which allows us to take account of complex parasite phenomena in the selected components (on the condition that the models used are sufficiently close to the real components which they claim to represent).

6 EDA: Electronics Design Automation.

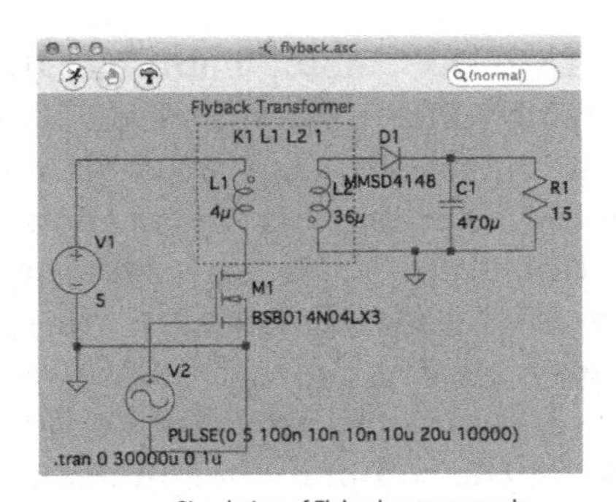

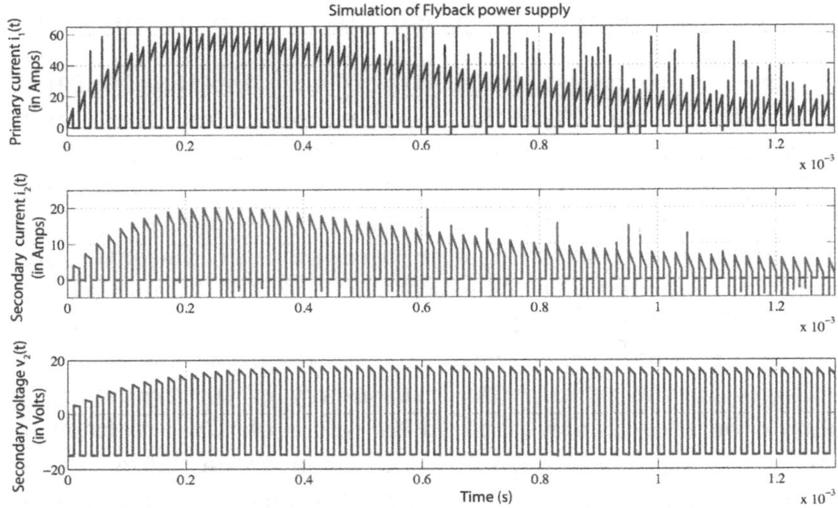

Figure 5.16. *Simulation of the Flyback power supply using LTSpice IV*

Based on this simulation, correct operation of the power supply may be validated using a number of measurements. First, we may verify the waveform of the transistor gate control signal. This waveform is shown in Figure 5.18. We see that this signal (denoted as Vgate) has an amplitude of 4 V (2 divisions, to give a caliber of 2 V/div.). It takes a quasi-perfect rectangular form, with rapid switching and no crossing of limits, due to a driver (Microchip MCP1416) placed between the microcontroller output point and the resistance connected to the transistor grid. This driver is able to deliver a current of 1.5 A to the transistor grid; as we have a gate resistance of 4.7 Ω (resistance specified in technical characteristics: the internal resistance when accessing the MOSFET gate should be added to this value), the driver is perfectly suited to our application.

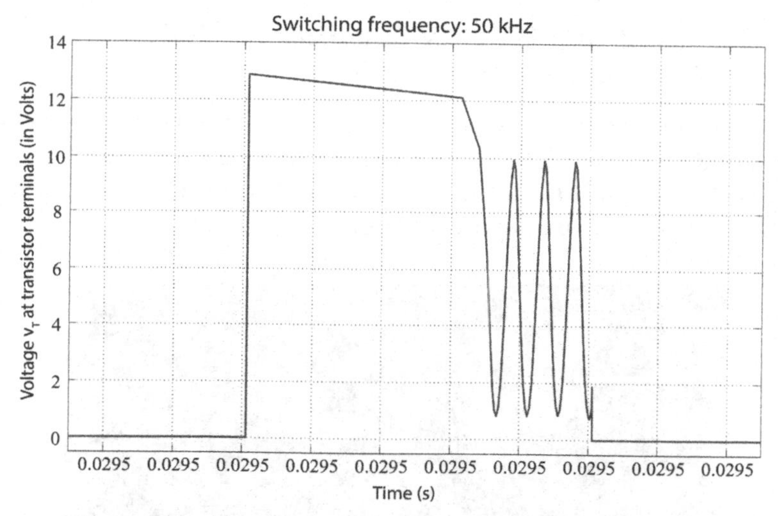

Figure 5.17. *Voltage waveform at the transistor terminals (discontinuous mode)*

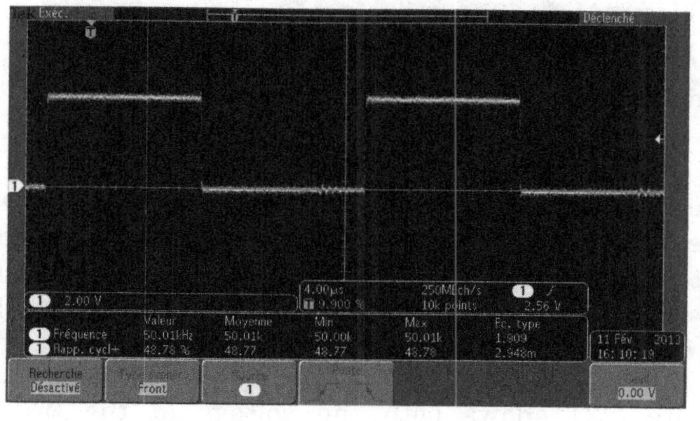

Figure 5.18. *Oscilloscope screenshot (Vgate). For a color version of the figure, see www.iste.co.uk/patin/power3.zip*

Figure 5.19 shows the waveform of the voltage at the transistor terminals (denoted as v_T): this is characteristic of discontinuous conduction. The theoretical values are respected (using a caliber of 5 V/div. for the graph):

– voltage zero during the transistor conduction phase;

– voltage of 10 V (5 V+15 V/3) during the diode conduction phase (if the output voltage is 15 V);

– average voltage equal to 5 V when the transistor and the diode are switched off (oscillations are present, as in the SPICE simulation, but with more damping, as losses are under-evaluated in the simulation).

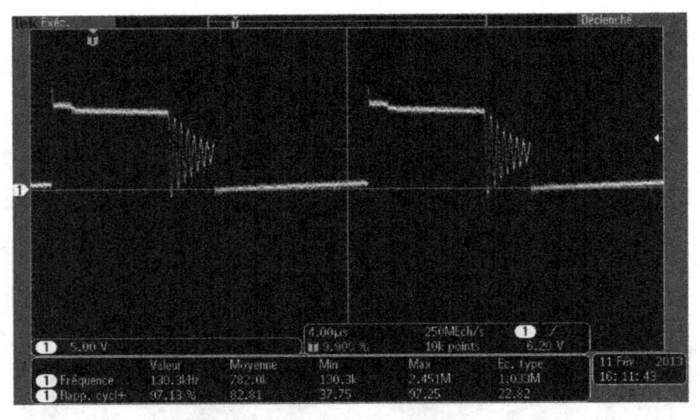

Figure 5.19. *Oscilloscope screenshot (v_T). For a color version of the figure, see www.iste.co.uk / patin / power3.zip*

The secondary voltage of the transformer is shown in Figure 5.20. As this is the output voltage of a transformer (denoted as v_2), we may verify that this conforms perfectly (with the turns ratio $m = 3$) to the primary voltage, subtracting the continuous component.

Figure 5.21 shows both the voltage at the transistor terminals and the primary current (measured using a Rogowski coil – so without the continuous component). We notice that the current evolves in a linear manner as a function of time when the transistor is on, and that it starts this phase with a value of zero. This corroborates our hypothesis for the discontinuous operation mode, with the voltage waveforms shown in Figures 5.19 and 5.20.

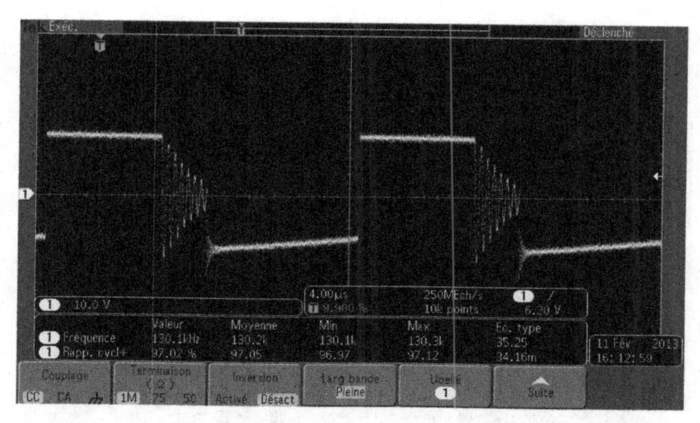

Figure 5.20. *Oscilloscope screenshot(v_2). For a color version of the figure, see www.iste.co.uk/patin/power3.zip*

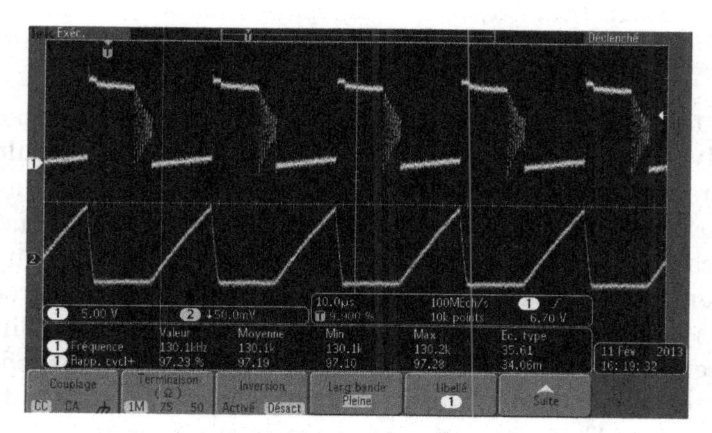

Figure 5.21. *Oscilloscope screenshot (top: v_T, bottom: i_1 – AC component alone). For a color version of the figure, see www.iste.co.uk/patin/power3.zip*

One final point needs to be tested in order to verify the correct operation of the system: the regulation of the output voltage (i.e. closed-loop operation). To do this, two separate tests are carried out (successively):

– switching on the power supply;

– observing the voltage regulation following a sudden load variation.

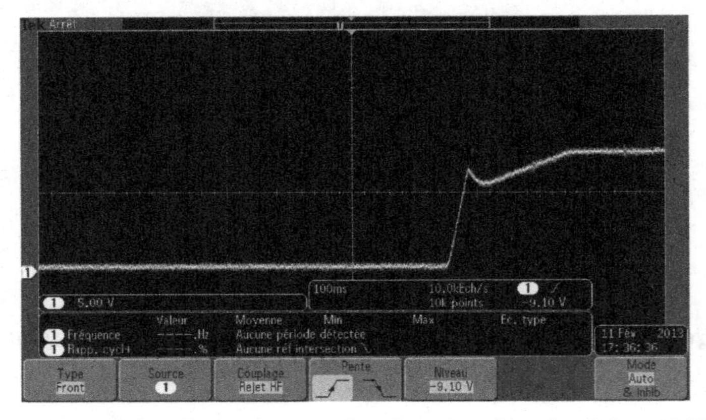

Figure 5.22. *Oscilliscope screenshot (starting from load $R = 30\,\Omega$). For a color version of the figure, see www.iste.co.uk/patin/power3.zip*

In the first case (shown in Figure 5.22), we start the power supply with a load equal to $30\,\Omega$ (note: open start-up would be the most destabilizing case for the system). To avoid excessive current draw, we generally carry out a soft start; in this case, this consists of applying a ramped voltage output reference value. The oscilloscope reading for this process shows that the response is characteristic of a non-linear device: note that the equivalent models established in Chapter 4 with the Middlebrook approach are deduced from "small signal" variations, and a power-up from 0 to $15\,\text{V}$ in this way lies outside of the modeling framework. This means that the system dynamics may differ considerably from theoretical predictions made on the basis of a linear model.

The second test is more compatible with a "small signal" modeling (resistor step variation) in that we wish to evaluate the response of the regulated power supply to a load impact, while the initial voltage is already equal to that of the reference value. Figure 5.23 shows the system response to an increase in the current draw from the load, caused by a reduction in the resistance used as a load (varying from 60 to 30 Ω). The oscilloscope reading shows a voltage drop of 3 V at the moment of impact, and that the voltage was re-established after around 50 ms. PI controller tuning does not offer particularly strong performances; however, the main aim of this is to facilitate observation of the impact of disturbances, rather than to provide optimal rejection of to disturbances by means of regulation.

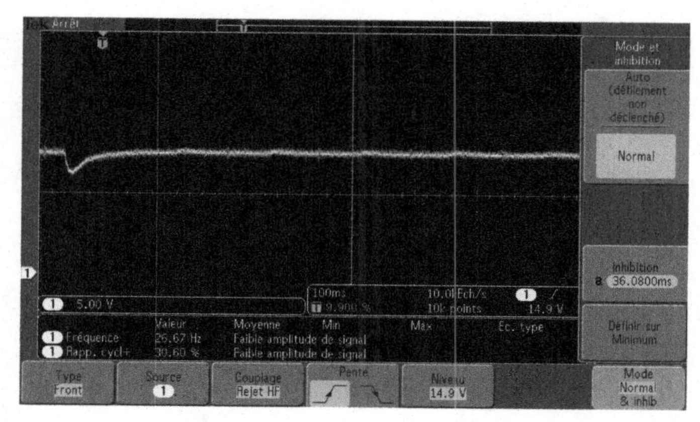

Figure 5.23. *Oscilloscope screenshot (impact of a variation from $R = 60\Omega$ to $R = 30\Omega$). For a color version of the figure, see www.iste.co.uk/patin/power3.zip*

Appendix 1

Formulas for Electrical Engineering and Electromagnetism

A1.1. Sinusoidal quantities

A1.1.1. *Scalar signals*

A1.1.1.1. *Definitions*

Sinusoidal waveforms are extremely widespread in electrical engineering, both for voltages and for currents. In this case, we will consider a generic signal of the form:

$$x(t) = X_{\max} \cos(\omega t - \varphi) \qquad [A1.1]$$

This real signal is associated with an equivalent complex signal:

$$\underline{x}(t) = X_{\max}.\mathrm{e}^{j(\omega t - \varphi)} \qquad [A1.2]$$

This vector may be represented in the complex plane. We obtain a circular trajectory of radius $X_{\max}$ with a vector rotating at a constant speed ω in a counterclockwise direction. This representation (which is widespread in electrical engineering) is known as a Fresnel diagram (or, more simply, a vector diagram).

REMARK A1.1.– Derivation and integration calculations are greatly simplified in the complex plane, as they are replaced, respectively, by multiplying or dividing by $j\omega$. To return to the real domain, we simply take the real part of the corresponding complex signal: $x(t) = \mathfrak{Re}\left[\underline{x}\,(t)\right]$.

The rotating component $e^{j\omega t}$ of the complex vectors is meaningless when studying linear circuits; the *amplitudes* and the relative phases between the different quantities under study are the only important elements. Note that an absolute phase for a sinusoidal value would be meaningless; the choice of a reference value of the form $X_{\text{ref}}.\cos(\omega.t)$, associated with the vector $X_{\text{ref}}.e^{j\omega t}$, is purely arbitrary.

Complex vectors are also often represented (in the literature) using the RMS value of the real value in question as the modulus, and not the real amplitude.

A1.1.1.2. *Trigonometric formulas*

When making calculations using complex values, we need Euler's formulas:

$$\begin{cases} \cos\theta = \frac{e^{j\theta}+e^{-j\theta}}{2} \\ \sin\theta = \frac{e^{j\theta}-e^{-j\theta}}{2j} \end{cases} \qquad \text{[A1.3]}$$

These two formulas can be used to give the four basic trigonometric formulas used in electrical engineering:

$$\begin{cases} \cos\left(a+b\right) = \cos a \cos b - \sin a \sin b \\ \cos\left(a-b\right) = \cos a \cos b + \sin a \sin b \\ \sin\left(a+b\right) = \sin a \cos b + \cos a \sin b \\ \sin\left(a-b\right) = \sin a \cos b - \cos a \sin b \end{cases} \qquad \text{[A1.4]}$$

These four equations allow us to establish four further equations:

$$\cos a \cos b = \frac{1}{2}\left(\cos\left(a+b\right) + \cos\left(a-b\right)\right) \qquad \text{[A1.5]}$$

$$\begin{cases} \cos a \cos b = \frac{1}{2}\left(\cos(a+b) + \cos(a-b)\right) \\ \sin a \sin b = \frac{1}{2}\left(\cos(a-b) - \cos(a-b)\right) \\ \sin a \cos b = \frac{1}{2}\left(\sin(a+b) + \sin(a-b)\right) \\ \cos a \sin b = \frac{1}{2}\left(\sin(a+b) - \sin(a-b)\right) \end{cases} \quad [A1.6]$$

A1.1.2. *Vector signals (three-phase context)*

A1.1.2.1. *Reference frame* (a, b, c)

Three-phase systems are very much common in electrical engineering, particularly balanced three-phase systems. A vector $(\mathbf{x}_3) = (x_a, x_b, x_c)^t$ with three balanced components is therefore expressed as:

$$(\mathbf{x}_3) = X_{\max} \begin{pmatrix} \cos\theta \\ \cos\left(\theta - \frac{2\pi}{3}\right) \\ \cos\left(\theta + \frac{2\pi}{3}\right) \end{pmatrix} \quad \text{where } \theta = \omega.t + \phi_0 \quad [A1.7]$$

in the case of a direct system, or:

$$(\mathbf{x}_3) = X_{\max} \begin{pmatrix} \cos\theta \\ \cos\left(\theta + \frac{2\pi}{3}\right) \\ \cos\left(\theta - \frac{2\pi}{3}\right) \end{pmatrix} \quad \text{where } \theta = \omega.t + \phi_0 \quad [A1.8]$$

in the inverse case.

DEFINITION A1.1.– A balanced three-phase system is thus made up of three sinusoids of the same amplitude and same frequency, with a phase deviation of $\frac{2\pi}{3}$.

A direct three-phase system is characterized by the fact that, taking phase 1 as a reference point (i.e. first component), the second component has a delay of 120° (in a balanced situation) and the third component presents a delay of 120° in relation to the second component.

An inverse three-phase system is characterized by the fact that, taking phase 1 as a reference point (i.e. first component), the third component has a delay of 120° (in a balanced situation) and the second component presents a delay of 120° in relation to the third component. A direct system can be converted into an inverse system (and vice versa) by permutations of two components.

A1.1.2.2. *Three-phase to two-phase transformation* (α, β)

It is important to note that a balanced three-phase system (whether direct or inverse) presents an important property in that the sum of the components is null:

$$x_a + x_b + x_c = 0 \qquad \text{[A1.9]}$$

This sum is classically referred to as the zero sequence component (denoted as x_0). A balanced three-phase system is therefore not linearly independent in that, given two of the components, we may calculate the value of the third component. It is therefore possible to propose a three-phase to two-phase transformation without any information loss. The simplest transformation, known as the Clarke (abc-to-$\alpha\beta$) transformation, allows us to associate an initial vector $(\mathbf{x}_3) = (x_a, x_b, x_c)^t$ with an equivalent two-phase vector $(\mathbf{x}_{\alpha\beta}) = (\mathbf{x}_2) = (x_\alpha, x_\beta)^t$ using components of the same amplitude as those in the initial vector. This operation introduces the Clarke matrix C_{32}:

$$X_{\max} \begin{pmatrix} \cos\theta \\ \cos\left(\theta + \frac{2\pi}{3}\right) \\ \cos\left(\theta - \frac{2\pi}{3}\right) \end{pmatrix} = X_{\max}. \underbrace{\begin{pmatrix} 1 & 0 \\ -1/2 & \sqrt{3}/2 \\ -1/2 & -\sqrt{3}/2 \end{pmatrix}}_{C_{32}} . \begin{pmatrix} \cos\theta \\ \sin\theta \end{pmatrix} \qquad \text{[A1.10]}$$

This gives the following direct transformation:

$$(\mathbf{x}_3) \triangleq C_{32}.(\mathbf{x}_2) \qquad \text{[A1.11]}$$

The Clarke transformation may be extended by taking account of the zero sequence component x_0, presented in [A1.9]:

$$(\mathbf{x}_3) \triangleq C_{32}.(\mathbf{x}_2) + C_{31}.x_0 \quad \text{[A1.12]}$$

with:

$$C_{31} = \begin{pmatrix} 1 \\ 1 \\ 1 \end{pmatrix} \quad \text{[A1.13]}$$

Noting certain properties of matrices C_{32} and C_{31}:

$$\begin{aligned} C_{32}^t C_{32} &= \tfrac{3}{2}\begin{pmatrix} 1\,0 \\ 0\,1 \end{pmatrix} \; ; \; C_{31}^t C_{31} = 3 \\ C_{32}^t C_{31} &= \begin{pmatrix} 0 \\ 0 \end{pmatrix} \; ; \; C_{31}^t C_{32} = \begin{pmatrix} 0\,0 \end{pmatrix} \end{aligned} \quad \text{[A1.14]}$$

we can establish the inverse transformation:

$$(\mathbf{x}_2) \triangleq \frac{2}{3} C_{32}^t.(\mathbf{x}_3) \quad \text{[A1.15]}$$

and:

$$x_0 \triangleq \frac{1}{3} C_{31}^t.(\mathbf{x}_3) \quad \text{[A1.16]}$$

A1.1.2.3. *Concordia variant*

A second three-phase to two-phase transformation is also widely used in the literature, with properties similar to those of the Clarke transformation. This variation does not retain the amplitudes of the transformed values, but allows us to retain powers. This operation is known as the Concordia transformation and is based on two matrices, denoted T_{32} and T_{31}, deduced from C_{32} and C_{31}:

$$T_{32} = \sqrt{\frac{2}{3}} C_{32} \; ; \; T_{31} = \frac{1}{\sqrt{3}} C_{31} \quad \text{[A1.17]}$$

The properties of these matrices are deduced from those established in [A1.14]:

$$T_{32}^t T_{32} = \begin{pmatrix} 1\,0 \\ 0\,1 \end{pmatrix} \; ; \; T_{31}^t T_{31} = 1$$
$$T_{32}^t T_{31} = \begin{pmatrix} 0 \\ 0 \end{pmatrix} \; ; \; T_{31}^t T_{32} = \begin{pmatrix} 0\,0 \end{pmatrix} \quad \text{[A1.18]}$$

This produces a direct transformation of the form:

$$(\mathbf{x}_3) \triangleq T_{32}.\,(\mathbf{x}_2) + T_{31}.x_0 \quad \text{[A1.19]}$$

with the following inverse transformation:

$$(\mathbf{x}_2) \triangleq T_{32}^t.\,(\mathbf{x}_3) \quad \text{[A1.20]}$$

and:

$$x_0 \triangleq T_{31}^t.\,(\mathbf{x}_3) \quad \text{[A1.21]}$$

A1.1.2.4. *Park transformation*

The Park (abc-to-dq) transformation consists of associating the Clarke (or Concordia) transformation with a rotation in the two-phase reference plane (α, β) onto a reference frame (d, q). This operation is carried out using the rotation matrix $P(\theta)$, defined as:

$$P(\theta) = \begin{pmatrix} \cos\theta & -\sin\theta \\ \sin\theta & \cos\theta \end{pmatrix} \quad \text{[A1.22]}$$

Thus, if we associate a vector $(\mathbf{x}_{dq}) = (x_d, x_q)^t$ with the initial two-phase vector $(\mathbf{x}_{\alpha\beta}) = (\mathbf{x}_2)$ (obtained from a Clarke or Concordia transformation), we obtain the following relationship:

$$(\mathbf{x}_{\alpha\beta}) = (\mathbf{x}_2) \triangleq P(\theta)\,.\,(\mathbf{x}_{dq}) \quad \text{[A1.23]}$$

The choice of a frame of reference involves the definition of angle θ, selected arbitrarily. Generally, the chosen reference frame is synchronous with the rotating values (sinusoidal components with an angular frequency ω), but this is not obligatory.

The following (non-exhaustive) list shows a number of properties of matrix $P(\theta)$:

$$P(0) = \begin{pmatrix} 1 & 0 \\ 0 & 1 \end{pmatrix} = \mathbb{I}_2 \; ; \; P\left(\tfrac{\pi}{2}\right) = \begin{pmatrix} 0 & -1 \\ 1 & 0 \end{pmatrix} = \mathbb{J}_2 \text{ such that } \mathbb{J}_2 = -\mathbb{I}_2 \qquad \text{[A1.24]}$$

$$P(\alpha + \beta) = P(\beta + \alpha) = P(\alpha).P(\beta) = P(\beta).P(\alpha) \qquad \text{[A1.25]}$$

$$P(\alpha)^{-1} = P(\alpha)^t = P(-\alpha) \qquad \text{[A1.26]}$$

$$\frac{d}{dt}\left[P(\alpha)\right] = \frac{d\alpha}{dt} \cdot P\left(\alpha + \frac{\pi}{2}\right) = \frac{d\alpha}{dt} \cdot P(\alpha) \cdot P\left(\frac{\pi}{2}\right) = \mathbb{J}_2 \frac{d\alpha}{dt} \cdot P(\alpha) \qquad \text{[A1.27]}$$

A1.1.2.5. *Phasers or complex vectors*

The matrix formalism of the Clarke, Concordia and Park transformations may be replaced by an equivalent complex representation. Evidently, a rotation of the frame of reference by angle θ may be obtained by using a complex coefficient $e^{j\theta}$ as easily as with a rotation matrix $P(\theta)$. To this end, we use a "phaser" $\underline{x}_s$ defined in a stationary frame of reference:

$$\underline{x}_s = x_\alpha + j.x_\beta \qquad \text{[A1.28]}$$

The phaser is also defined in a rotating frame ($\underline{x}_r$):

$$\underline{x}_r = x_d + j.x_q \qquad \text{[A1.29]}$$

Note that these complex representations may be obtained using matrix transformations. The real transformations seen in the previous sections each have an equivalent complex transformation, as shown in Table A1.1.

Real transformation	Complex transformation
Clarke	Fortescue
Concordia	Lyon
Park	Ku

Table A1.1. *Correspondence between real and complex transformations (names)*

A1.2. General characteristics of signals in electrical engineering

This section presents the formulas used for calculating the *general characteristics of periodic signals* traditionally encountered in electrical engineering. However, it does not cover formulas related to spectral analysis, which are covered in Appendix 2 of Volumes 2 and 4 ([PAT 15b] and [PAT 15c]) of this series.

In this section, we will therefore cover the formulas used to calculate the average and RMS values of given quantities, applied to two widespread signal types: sinusoids and the asymmetric square signal of duty ratio α.

A1.2.1. *Average value*

A1.2.1.1. *General definition*

The average value $\langle x \rangle$ of a T-periodic signal $x(t)$ is defined generally by the integral:

$$\langle x \rangle = \frac{1}{T} \int_0^T x(t).dt \qquad \text{[A1.30]}$$

REMARK A1.2.– In this case, the integration limits are chosen arbitrarily. Only the interval between the two limits is important, and it must be equal to T.

A1.2.1.2. *Sinusoids*

In the case of sinusoids, we evidently obtain an average value of zero.

A1.2.1.3. *Asymmetric square*

The T-periodic asymmetric square $x(t)$ studied here has a certain value X_0 during a period αT, then 0 for the rest of the period. We can therefore write the average value $\langle x \rangle$ directly:

$$\langle x \rangle = \frac{1}{T} \int_0^T x(t).dt = \frac{1}{T} \int_0^{\alpha T} X_0.dt = \alpha.X_0 \qquad \text{[A1.31]}$$

A1.2.2. *RMS value*

A1.2.2.1. *General definition*

The RMS value X_{rms} of a T-periodic signal $x(t)$ is defined generally by the integral:

$$X_{\text{rms}} = \sqrt{\frac{1}{T} \int_0^T x^2(t).dt} \qquad \text{[A1.32]}$$

REMARK A1.3.– When calculating the average value, the integration limits are chosen arbitrarily. Only the interval between the two limits is important, and it must be equal to T.

A1.2.2.2. *Sinusoids*

For a sinusoid of amplitude $X_{\max}$, the RMS value is $X_{\text{rms}} = \frac{X_{\max}}{\sqrt{2}}$.

A1.2.2.3. *Asymmetric square*

The T-periodic asymmetric square $x(t)$ defined in section A1.2.1 presents an RMS value expressed as:

$$X_{\text{rms}} = \sqrt{\frac{1}{T} \int_0^{\alpha T} X_0^2.dt} = \sqrt{\alpha}.X_0 \qquad \text{[A1.33]}$$

A1.3. Energy and power

A1.3.1. *Energy*

In mechanics, energy is obtained by the operation of a force over a certain distance. In electrical engineering, this term corresponds to the movement of a charge following a variation in electrical potential. In particle physics, a unit known as an electron-volt (eV) is used for energy values at the atomic level. The energy formulas used in power electronics (expressed in Joules (J)) correspond to the energy stored in an inductance or a capacitor.

In an inductance, the energy E_L (magnetic energy) depends on the current I and the inductance L:

$$E_L = \frac{1}{2}LI^2 \qquad [A1.34]$$

For a capacitor, the energy E_C (electrostatic energy) depends on the voltage V and the capacitance C:

$$E_C = \frac{1}{2}CV^2 \qquad [A1.35]$$

A1.3.2. *Instantaneous power*

The instantaneous power $p(t)$ given – or provided to – the dipole is linked, according to the passive sign convention (PSC), to the voltage $v(t)$ at its terminals and the current $i(t)$ passing through it as follows:

$$p(t) = v(t).i(t) \qquad [A1.36]$$

This power is defined in watts (W). It is linked to the energy consumed E (in J) between two instants t_1 and t_2 by the following integral:

$$E = \int_{t_1}^{t_2} p(t).dt \qquad [A1.37]$$

The instantaneous power $p(t)$ is connected to the variation in energy $e(t)$ which can also be defined (up to an additive constant) as a function of time. In this case, we obtain:

$$p(t) = \frac{de(t)}{dt} \qquad [A1.38]$$

A1.3.3. *Average power*

As for any T-periodic signal, the average power P is obtained using the following formula:

$$P = \frac{1}{T}\int_0^T p(t).dt = \frac{1}{T}\int_0^T v(t).i(t).dt \qquad [A1.39]$$

In the case of a resistive charge R, we can establish the following relationship (Ohm's law):

$$v(t) = R.i(t) \qquad [A1.40]$$

This allows us to formulate two possible expressions for this power:

$$P = \frac{R}{T}\int_0^T i^2(t).dt = R.I_{\text{rms}}^2 \qquad [A1.41]$$

and:

$$P = \frac{1}{RT}\int_0^T v^2(t).dt = \frac{V_{\text{rms}}^2}{R} \qquad [A1.42]$$

where V_{rms} and I_{rms} are the RMS values of the voltage and the current, respectively.

A1.3.4. *Sinusoidal mode*

A1.3.4.1. *Single phase*

In single phase sinusoidal operating mode, we can, generally speaking, consider a voltage $v(t)$ of the form:

$$v(t) = V_{\mathrm{rms}}\sqrt{2}\cos(\omega t) \qquad [A1.43]$$

as the phase reference, with a current, with a phase deviation angle φ (the lag in relation to the voltage), expressed as:

$$i(t) = I_{\mathrm{rms}}\sqrt{2}\cos(\omega t - \varphi) \qquad [A1.44]$$

Calculating the instantaneous power obtained using these two values, we obtain:

$$p(t) = V_{\mathrm{rms}}I_{\mathrm{rms}}\left(\cos(2\omega t - \varphi) + \cos\varphi\right) \qquad [A1.45]$$

We thus obtain two terms:

– a constant term, which is, evidently, the average power, referred to in this context as active power;

– a variable term, with an angular frequency of 2ω, known as fluctuating power.

The first interesting result is, therefore, the expression of the average (active) power P:

$$P = V_{\mathrm{rms}}I_{\mathrm{rms}}\cos\varphi \qquad [A1.46]$$

In terms of voltage dimensioning (thickness of insulation) and current dimensioning (cross-section of conductors) of equipment, the real power value used for design purposes is known as the apparent power S , and is obtained by directly multiplying the RMS voltage value by the RMS current value:

$$S = V_{\mathrm{rms}}I_{\mathrm{rms}} \qquad [A1.47]$$

To emphasize the "fictional" character of this power, it is not given in W, but in volt-amperes (VA).

In electrical engineering, we then use the notion of *reactive power* Q, which allows us to establish a connection between the active power P and the apparent power S. This is expressed as:

$$Q = V_{\text{rms}} I_{\text{rms}} \sin \varphi \qquad \text{[A1.48]}$$

The connection between P, Q and S is thus:

$$S^2 = P^2 + Q^2 \qquad \text{[A1.49]}$$

As in the case of apparent power, this power value is fictional; it is not measured in W, or in VA, but rather in volt ampere reactive (VAR).

REMARK A1.4.– Equation [A1.49] is only valid if the voltage *and* the current are sinusoidal. In non-sinusoidal mode, we introduce an additional power, denoted D, known as the deformed power. This is used to establish a new equation as follows:

$$S^2 = P^2 + Q^2 + D^2 \qquad \text{[A1.50]}$$

The instantaneous power is always positive (respectively, negative) when $\varphi = 0°$ (respectively, $\varphi = 180°$), but if φ takes a different value, $p(t)$ cancels out, changing the sign. In these conditions, the direction of transfer of electronic energy between the source and the load is reversed.

A1.3.4.2. *Three phase*

In a three-phase context, using the "voltage" vector (v_3) as a point of reference, and more specifically as the first component, we take (based on the hypothesis of a direct balanced system):

$$(\mathbf{v}_3) = V_{\text{rms}}\sqrt{2}\begin{pmatrix} \cos(\omega t) \\ \cos\left(\omega t - \frac{2\pi}{3}\right) \\ \cos\left(\omega t + \frac{2\pi}{3}\right) \end{pmatrix} \quad \text{[A1.51]}$$

From this, we deduce the "current" vector $(\mathbf{i}_3)$, with a lag in each component when compared to the corresponding components in $(\mathbf{v}_3)$:

$$(\mathbf{i}_3) = I_{\text{rms}}\sqrt{2}\begin{pmatrix} \cos(\omega t - \varphi) \\ \cos\left(\omega t - \frac{2\pi}{3} - \varphi\right) \\ \cos\left(\omega t + \frac{2\pi}{3} - \varphi\right) \end{pmatrix} \quad \text{[A1.52]}$$

A matrix formalism may be used to obtain the expression of the instantaneous power $p(t)$:

$$p(t) = (\mathbf{v}_3)^t \cdot (\mathbf{i}_3) \quad \text{[A1.53]}$$

In this case, the Park factorization of the "voltage and current" vectors is particularly effective:

$$(\mathbf{v}_3) = V_{\text{rms}}\sqrt{2}.C_{32}\begin{pmatrix} \cos(\omega t) \\ \sin(\omega t) \end{pmatrix}$$

$$= V_{\text{rms}}\sqrt{2}.C_{32}.P(\omega t) \cdot \begin{pmatrix} 1 \\ 0 \end{pmatrix} \quad \text{[A1.54]}$$

$$(\mathbf{i}_3) = I_{\text{rms}}\sqrt{2}.C_{32}\begin{pmatrix} \cos(\omega t - \varphi) \\ \sin(\omega t - \varphi) \end{pmatrix}$$

$$= I_{\text{rms}}\sqrt{2}.C_{32}.P(\omega t - \varphi) \cdot \begin{pmatrix} 1 \\ 0 \end{pmatrix} \quad \text{[A1.55]}$$

Hence:

$$p(t) = 2V_{\text{rms}}.I_{\text{rms}} \begin{pmatrix} 1 & 0 \end{pmatrix} .P(-\omega t).C_{32}^{t}.C_{32}.P(\omega t - \varphi).\begin{pmatrix} 1 \\ 0 \end{pmatrix} \quad [A1.56]$$

After simplification, this gives us:

$$p(t) = 3V_{\text{rms}}.I_{\text{rms}} \cos\varphi \quad [A1.57]$$

Note that we obtain the instantaneous power, and not an average value. This highlights a notable property of three-phase systems: there is no globally fluctuating power in this configuration.

The active power P is therefore defined as follows:

$$P = p(t) = 3V_{\text{rms}}.I_{\text{rms}} \cos\varphi \quad [A1.58]$$

The notions of reactive power Q and apparent power S are also used in three-phase contexts, with the following expressions:

$$\begin{cases} Q = 3V_{\text{rms}}.I_{\text{rms}} \sin\varphi \\ S = 3V_{\text{rms}}.I_{\text{rms}} \end{cases} \quad [A1.59]$$

Relationship [A1.49] is therefore still valid in a three-phase context:

$$S^2 = P^2 + Q^2 \quad [A1.60]$$

Note that variants exist, notably where the notion of line-to-line voltage is used. Voltage V_{rms} is the RMS *line-to-neutral voltage* (i.e. between the phase and the neutral); the neutral is not always accessible, so the notion of line-to-line voltage is often preferred , with an RMS voltage, denoted U_{rms}. In the case of a balanced three-phase system, the relationship between the RMS line-to-neutral and line-to-line voltage is:

$$U_{\text{rms}} = \sqrt{3}V_{\text{rms}} \quad \text{[A1.61]}$$

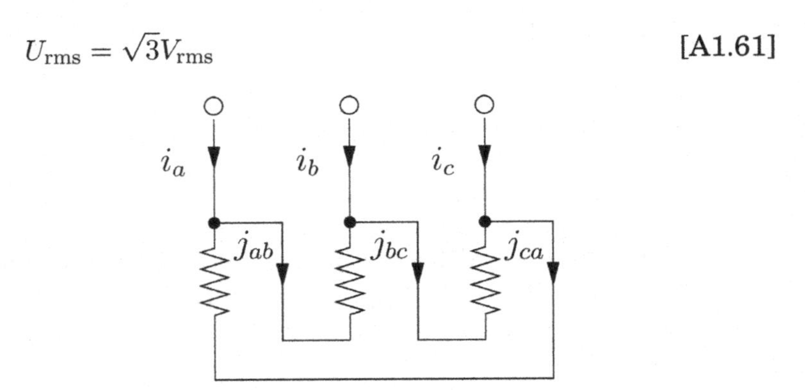

Figure A1.1. *Line and branch currents for a triangular connection*

A second point, which may lead to a different formulation of expression [A1.58], is concerned with currents. Generally speaking, we always have access to *line currents*, and thus to the RMS value I_{rms}. A second type of current can appear when using a load with a triangle connection (see Figure A1.1): this branch current presents an RMS value J_{rms} with the following expression as a function of I_{rms}:

$$J_{\text{rms}} = \frac{I_{\text{rms}}}{\sqrt{3}} \quad \text{[A1.62]}$$

A1.4. Mathematics for electromagnetism

A1.4.1. *The Green–Ostrogradsky theorem*

The Green–Ostrogradsky theorem (also known as the flux–divergence theorem) establishes a connection between the integral of the divergence of a field with vector **E** in a volume Ω and the integral of the flux of **E** on the closed surface $\partial\Omega$ delimiting the volume Ω:

$$\iiint_{\Omega} \text{div}\mathbf{E}.d\omega = \oiint_{\partial\Omega} \mathbf{E} \cdot d\mathbf{s} \qquad \text{[A1.63]}$$

where $d\omega$ is a volume element, while $d\mathbf{s}$ is a normal vector[1] with a surface element (infinitesimal) ds of the complete surface $\partial\Omega$.

A1.4.2. *Stokes–Ampère theorem*

The Stokes–Ampère theorem establishes a connection between the the flux curl of the magnetic field **H** on a surface Σ and the integral of the circulation of **H** along the closed contour $\partial\Sigma$ delimiting surface Σ:

$$\iint_{\Sigma} \mathbf{curl}\,\mathbf{H} \cdot d\mathbf{s} = \oint_{\partial\Sigma} \mathbf{H} \cdot d\mathbf{l} \qquad \text{[A1.64]}$$

where $d\mathbf{s}$ is a normal vector[2] with a surface element (infinitesimal) ds of the complete surface Σ. Element $d\mathbf{l}$ is a vector (whose norm is dl) tangent to the closed contour $\partial\Sigma$.

A1.4.3. *Differential and referential operators*

The definition of the differential operators used in electromagnetism (primarily **grad**, div and **curl**) is dependent on the chosen frame of reference. Using the Cartesian

1 Oriented toward the outside of volume Ω.

2 Oriented in accordance with the right-hand rule as a function of the choice of orientation of contour $\partial\Sigma$.

coordinate system, the nabla operator (vector), ∇, allows us to easily write these operators as:

$$\nabla = \begin{pmatrix} \frac{\partial}{\partial x} \\ \frac{\partial}{\partial y} \\ \frac{\partial}{\partial z} \end{pmatrix} \quad \text{[A1.65]}$$

and we know that:

$$\begin{cases} \mathbf{grad}\, V = \nabla V \\ \text{div}\, \mathbf{E} = \nabla \cdot \mathbf{E} \\ \mathbf{curl}\, \mathbf{H} = \nabla \times \mathbf{H} \end{cases} \quad \text{[A1.66]}$$

where the symbol "·" is the scalar product and "×" is the vector product.

If we want to write these operators using spherical or cylindrical coordinates, the ∇ operator is no longer suitable; in these cases, it is better to use intrinsic definitions (which are independent of the chosen frame of reference). For the gradient, we have:

$$dV = (\mathbf{grad}\, V) \cdot d\mathbf{r} \quad \text{[A1.67]}$$

where dV is the exact total differential of V and dr is an infinitesimal shift (vector) away from the considered point in the space (defined by vector r from the origin of the reference frame).

For the "divergence" and "curl" operators, we simply use the two theorems presented in sections A1.4.1 and A1.4.2. First, we obtain:

$$d\phi = \text{div}\, \mathbf{E}.d\omega \quad \text{[A1.68]}$$

where $d\phi$ is the flux of **E** across the surface of the volume $d\omega$ under consideration.

We can then write:

$$dC = \mathbf{curl\,H} \cdot \mathrm{n}.dS \qquad [A1.69]$$

where dC is the circulation of field H along a closed contour enclosing a surface dS, and with an orientation allowing us to define a normal (unitary) vector n (in accordance with the right-hand rule).

Appendix 2

Technical Documentation for Components

This appendix includes full technical documentation for the MOSFET power transistor and Schottky diode used in the flyback power supply presented in Chapter 5. We have also included the documentation for the Microchip MCP1416 driver used to drive the transistor, and for the Vishay IL300 circuit, used to enable isolated measurement of the power supply output voltage; this element is essential to maintain the galvanic isolation acquired by the power structure on the command side (fed, in this case, from the "primary" side of the "transformer").

A2.1. MOSFET power transistor

The technical documentation for this component has been described in Figures A2.1–A2.9. The MOSFET power transistor STD17NF03 is supplied by ST Microelectronics (www.st.com).

STD17NF03L
STD17NF03L-1

N-channel 30V - 0.038Ω - 17A - DPAK/IPAK
STripFET™ II Power MOSFET

General features

Type	V_{DSS}	$R_{DS(on)}$	I_D
STD17NF03L-1	30V	<0.05Ω	17A
STD17NF03L	30V	<0.05Ω	17A

- Exceptional dv/dt capability
- Low gate charge at 100°C
- Application oriented characterization
- 100% avalanche tested

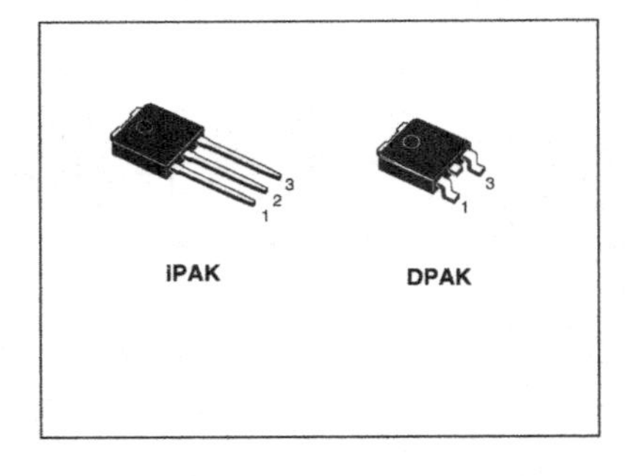

Description

This Power MOSFET is the latest development of STMicroelectronics unique "Single Feature Size™" strip-based process. The resulting transistor shows extremely high packing density for low on-resistance, rugged avalanche characteristics and less critical alignment steps therefore a remarkable manufacturing reproducibility.

Applications

- Switching application

Internal schematic diagram

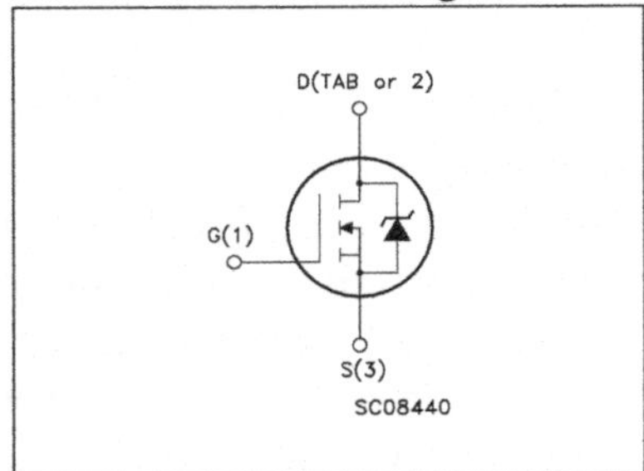

Order codes

Part number	Marking	Package	Packaging
STD17NF03L-1	D17NF03L@	IPAK	Tube
STD17NF03LT4	D17NF03L@	DPAK	Tape & reel

Figure A2.1. *MOSFET power transistor: ST-Microelectronics (Datasheet)*

1 Electrical ratings

Table 1. Absolute maximum ratings

Symbol	Parameter	Value	Unit
V_{DS}	Drain-source voltage (V_{GS} = 0)	30	V
V_{DGR}	Drain-gate voltage (R_{GS} = 20 kΩ)	30	V
V_{GS}	Gate- source voltage	± 16	V
I_D	Drain current (continuous) at T_C = 25°C	17	A
I_D	Drain current (continuous) at T_C = 100°C	12	A
I_{DM} (1)	Drain current (pulsed)	68	A
P_{tot}	Total dissipation at T_C = 25°C	30	W
	Derating Factor	0.2	W/°C
dv/dt (2)	Peak diode recovery avalanche energy	7	V/ns
E_{AS} (3)	Single pulse avalanche energy	200	mJ
T_{stg}	Storage temperature	-55 to 175	°C
T_j	Max. operating junction temperature		

1. Pulse width limited by safe operating area.
2. $I_{SD} \leq 17A$, $di/dt \leq 300A/\mu s$, $V_{DD} = V_{(BR)DSS}$, $T_j \leq T_{JMAX}$
3. Starting T_j = 25 °C, I_D = 8.5A, V_{DD} = 15V

Table 2. Thermal data

Rthj-case	Thermal resistance junction-case max	5.0	°C/W
Rthj-amb	Thermal resistance junction-to ambient max	100	°C/W
T_J	Maximum lead temperature for soldering purpose	275	°C

Figure A2.2. *MOSFET power transistor: ST-Microelectronics (Datasheet)*

2 Electrical characteristics

(T_{CASE}=25°C unless otherwise specified)

Table 3. On/off states

Symbol	Parameter	Test conditions	Min.	Typ.	Max.	Unit
$V_{(BR)DSS}$	Drain-source breakdown voltage	I_D = 250μA, V_{GS} =0	30			V
I_{DSS}	Zero gate voltage drain current (V_{GS} = 0)	V_{DS} = Max rating V_{DS} = Max rating, T_C = 125°C			1 10	μA μA
I_{GSS}	Gate-body leakage current (V_{DS} = 0)	V_{GS} = ± 16V			±100	nA
$V_{GS(th)}$	Gate threshold voltage	V_{DS} = V_{GS}, I_D = 250μA	1	1.5	2.2	V
$R_{DS(on)}$	Static drain-source on resistance	V_{GS} = 10V, I_D = 8.5A V_{GS} = 5V, I_D = 8.5A		0.038 0.045	0.05 0.06	Ω Ω

Table 4. Dynamic

Symbol	Parameter	Test conditions	Min.	Typ.	Max.	Unit
g_{fs} (1)	Forward transconductance	V_{DS} > $I_{D(on)}$ x $R_{DS(on)max}$, I_D =8.5A		12		S
C_{iss} C_{oss} C_{rss}	Input capacitance Output capacitance Reverse transfer capacitance	V_{DS} = 25V, f = 1MHz, V_{GS} = 0		320 155 28		pF pF pF
$t_{d(on)}$ t_r $t_{d(off)}$ t_f	Turn-on delay time Rise time Turn-off delay time Fall time	V_{DD} = 15V, I_D = 8.5A R_G = 4.7Ω V_{GS} = 5V (see *Figure 13*)		11 100 25 22		ns ns ns ns
Q_g Q_{gs} Q_{gd}	Total gate charge Gate-source charge Gate-drain charge	V_{DD} = 3024V, I_D = 17A, V_{GS} = 5V, R_G = 4.7Ω (see *Figure 14*)		4.8 2.25 1.7	6.5	nC nC nC

1. Pulsed: Pulse duration = 300 μs, duty cycle 1.5%.

Figure A2.3. *MOSFET power transistor: ST-Microelectronics (Datasheet)*

Table 5. Source drain diode

Symbol	Parameter	Test conditions	Min.	Typ.	Max.	Unit
I_{SD} I_{SDM} (1)	Source-drain current Source-drain current (pulsed)				22 88	A A
V_{SD} (2)	Forward on voltage	I_{SD} = 17A, V_{GS} = 0			1.5	V
t_{rr} Q_{rr} I_{RRM}	Reverse recovery time Reverse recovery charge Reverse recovery current	I_{SD} = 17A, di/dt = 100A/µs, V_{DD} = 15V, T_j = 150°C (see *Figure 15*)		28 18 1.3		ns nC A

1. Pulse width limited by safe operating area.
2. Pulsed: Pulse duration = 300 µs, duty cycle 1.5%

Figure A2.4. *MOSFET power transistor: ST-Microelectronics (Datasheet)*

2.1 Electrical characteristics (curves)

Figure 1. **Safe operating area** **Figure 2.** **Thermal impedance**

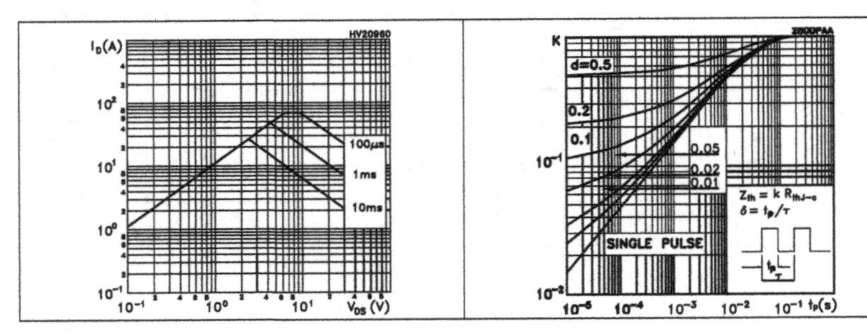

Figure 3. **Output characteristics** **Figure 4.** **Transfer characteristics**

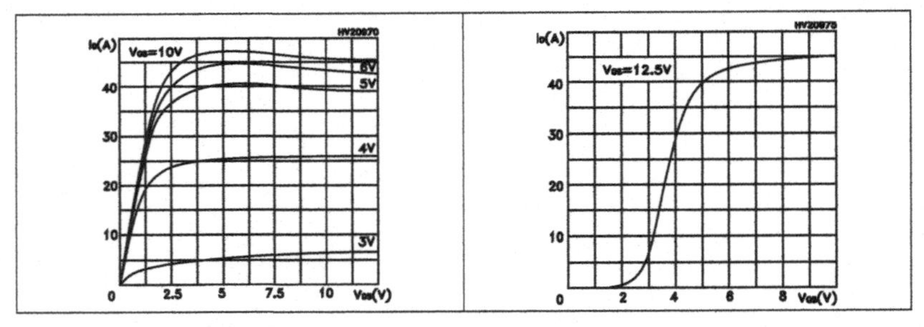

Figure 5. **Transconductance** **Figure 6.** **Static drain-source on resistance**

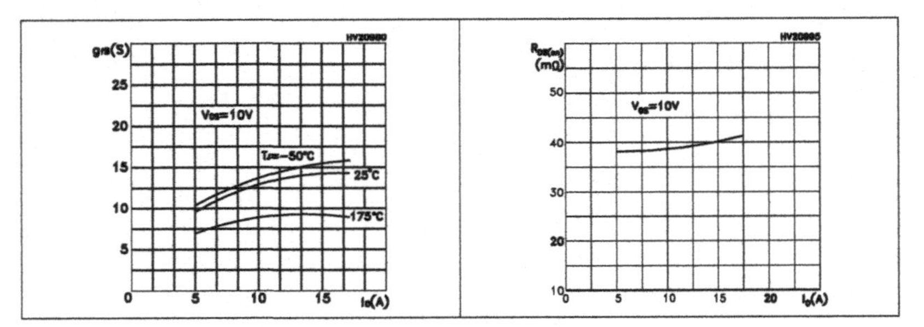

Figure A2.5. *MOSFET power transistor: ST-Microelectronics (Datasheet)*

STD17NF03L - STD17NF03L-1 **Electrical characteristics**

Figure 7. Gate charge vs. gate-source voltage **Figure 8. Capacitance variations**

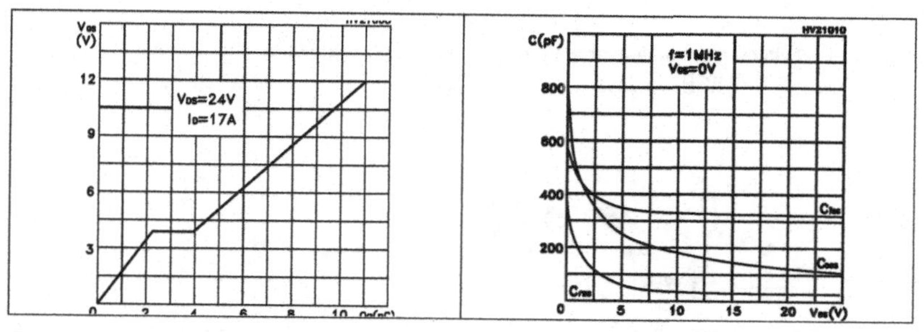

Figure 9. Normalized gate threshold voltage vs. temperature **Figure 10. Normalized on resistance vs. temperature**

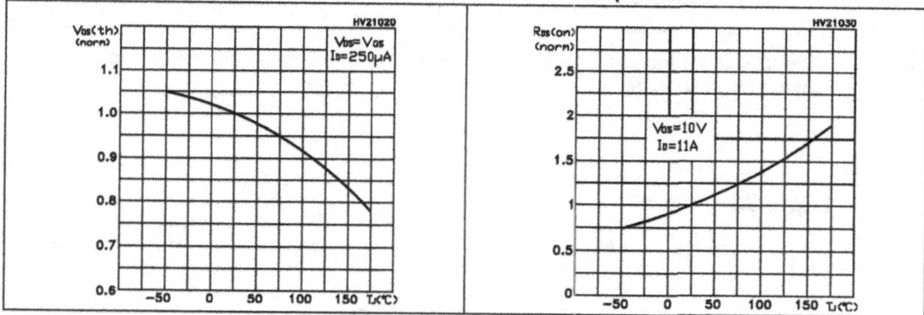

Figure 11. Source-drain diode forward characteristics **Figure 12. Normalized breakdown voltage vs. temperature**

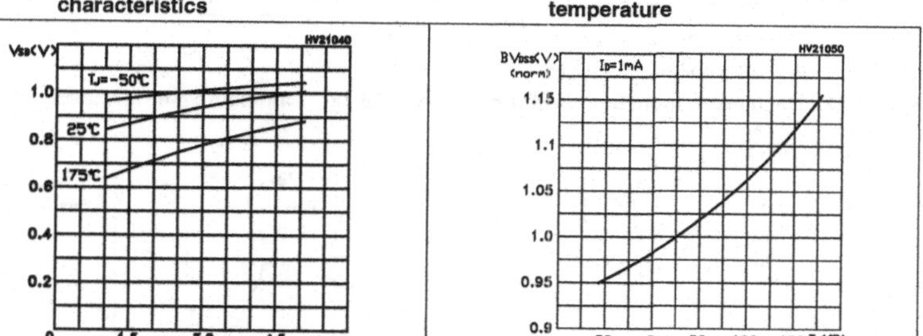

7/14

Figure A2.6. *MOSFET power transistor: ST-Microelectronics (Datasheet)*

3 Test circuit

Figure 13. Switching times test circuit for resistive load

Figure 14. Gate charge test circuit

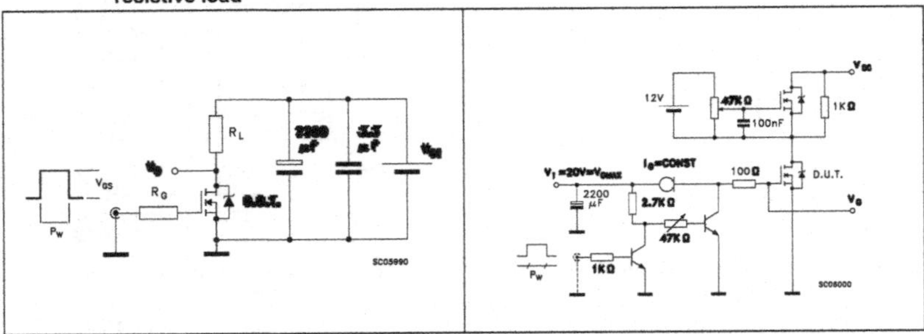

Figure 15. Test circuit for inductive load switching and diode recovery times

Figure 16. Unclamped Inductive load test circuit

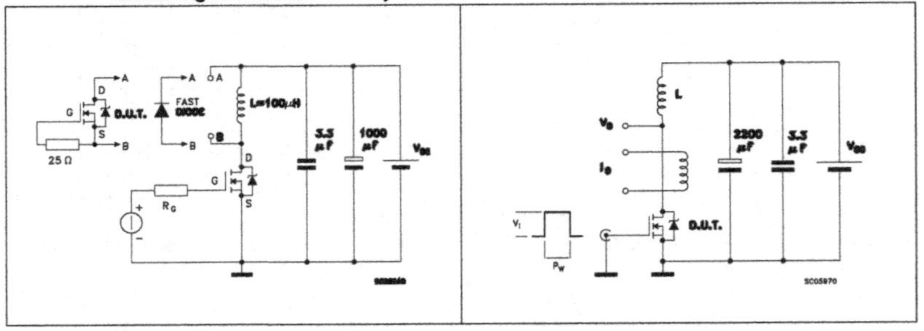

Figure 17. Unclamped inductive waveform

Figure 18. Switching time waveform

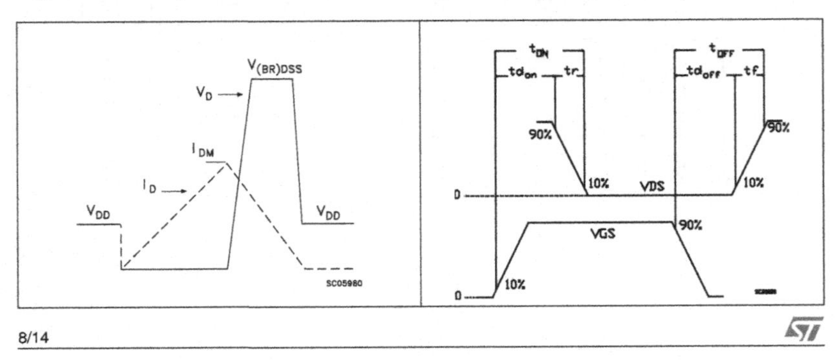

Figure A2.7. *MOSFET power transistor: ST-Microelectronics (Datasheet)*

DPAK MECHANICAL DATA						
DIM.	mm.			inch		
	MIN.	TYP	MAX.	MIN.	TYP.	MAX.
A	2.2		2.4	0.086		0.094
A1	0.9		1.1	0.035		0.043
A2	0.03		0.23	0.001		0.009
B	0.64		0.9	0.025		0.035
b4	5.2		5.4	0.204		0.212
C	0.45		0.6	0.017		0.023
C2	0.48		0.6	0.019		0.023
D	6		6.2	0.236		0.244
D1		5.1			0.200	
E	6.4		6.6	0.252		0.260
E1		4.7			0.185	
e		2.28			0.090	
e1	4.4		4.6	0.173		0.181
H	9.35		10.1	0.368		0.397
L	1			0.039		
(L1)		2.8			0.110	
L2		0.8			0.031	
L4	0.6		1	0.023		0.039
R		0.2			0.008	
V2	0°		8°	0°		8°

0068772-F

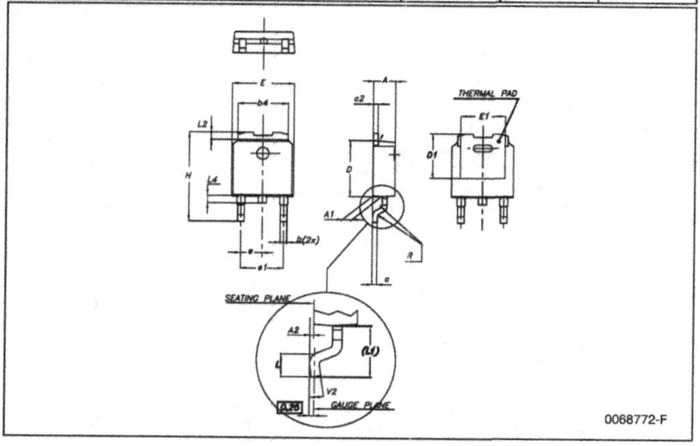

Figure A2.8. *MOSFET power transistor: ST-Microelectronics (Datasheet)*

5 Packing mechanical data

DPAK FOOTPRINT

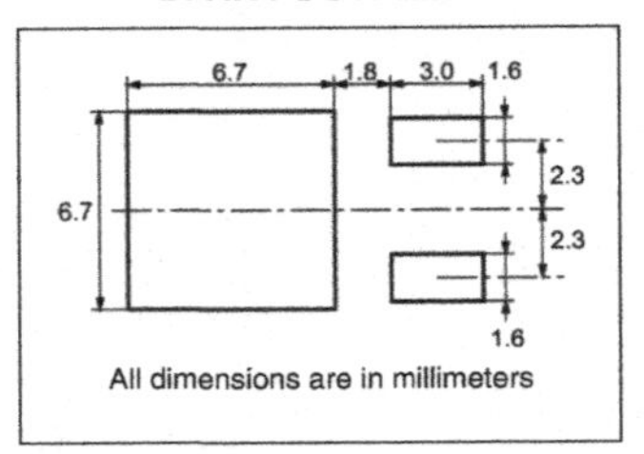

TAPE AND REEL SHIPMENT

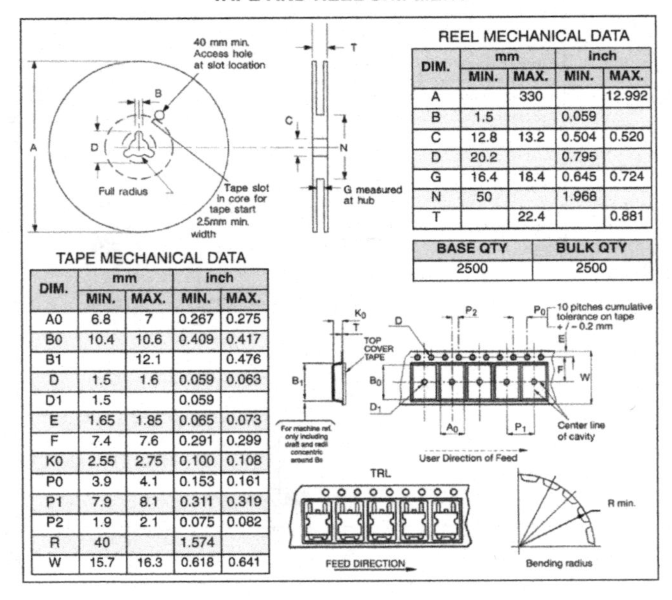

REEL MECHANICAL DATA

DIM.	mm		inch	
	MIN.	MAX.	MIN.	MAX.
A		330		12.992
B	1.5		0.059	
C	12.8	13.2	0.504	0.520
D	20.2		0.795	
G	16.4	18.4	0.645	0.724
N	50		1.968	
T		22.4		0.881

BASE QTY	BULK QTY
2500	2500

TAPE MECHANICAL DATA

DIM.	mm		inch	
	MIN.	MAX.	MIN.	MAX.
A0	6.8	7	0.267	0.275
B0	10.4	10.6	0.409	0.417
B1		12.1		0.476
D	1.5	1.6	0.059	0.063
D1	1.5		0.059	
E	1.65	1.85	0.065	0.073
F	7.4	7.6	0.291	0.299
K0	2.55	2.75	0.100	0.108
P0	3.9	4.1	0.153	0.161
P1	7.9	8.1	0.311	0.319
P2	1.9	2.1	0.075	0.082
R	40		1.574	
W	15.7	16.3	0.618	0.641

Figure A2.9. *MOSFET power transistor power transistor: ST-Microelectronics (Datasheet)*

A2.2. Schottky diode

The full documentation for the Schottky 240-13-F diode, manufactured by Diodes Incorporated (www.diodes.com), is shown below (Figures A2.10–A2.13).

B220/A - B260/A

2.0A SURFACE MOUNT SCHOTTKY BARRIER RECTIFIER

Features

- Guard Ring Die Construction for Transient Protection
- Ideally Suited for Automated Assembly
- Low Power Loss, High Efficiency
- Surge Overload Rating to 50A Peak
- For Use in Low Voltage, High Frequency Inverters, Free Wheeling, and Polarity Protection Application
- High Temperature Soldering: 260°C/10 Second at Terminal
- **Lead-Free Finish; RoHS Compliant (Notes 1 & 2)**
- **Halogen and Antimony Free. "Green" Device (Note 3)**
- **Qualified to AEC-Q101 Standards for High Reliability**

Mechanical Data

- Case: SMA/SMB
- Case Material: Molded Plastic. UL Flammability Classification Rating 94V-0
- Moisture Sensitivity: Level 1 per J-STD-020
- Terminals: Lead Free Plating (Matte Tin Finish). Solderable per MIL-STD-202, Method 208 e3
- Polarity: Cathode Band or Cathode Notch
- Weight: SMA 0.064 grams (Approximate)
  SMB 0.093 grams (Approximate)

SMA/SMB

Top View

Bottom View

Ordering Information (Note 4)

Part Number	Qualification	Case	Packaging
B2xxA-13-F	Standard	SMA	5000/Tape & Reel
B2xx-13-F	Standard	SMB	3000/Tape & Reel
B250Q-13	Automotive	SMB	3000/Tape & Reel
B240AQ-13-F	Automotive	SMA	5000/Tape & Reel
B240Q-13-F	Automotive	SMB	3000/Tape & Reel

* x = Device type, e.g. B260A-13-F (SMA package); B240-13-F (SMB package).

Notes:
1. EU Directive 2002/95/EC (RoHS) & 2011/65/EU (RoHS 2) compliant. All applicable RoHS exemptions applied.
2. See http://www.diodes.com/quality/lead_free.html for more information about Diodes Incorporated's definitions of Halogen- and Antimony-free, "Green" and Lead-free.
3. Halogen- and Antimony-free "Green" products are defined as those which contain <900ppm bromine, <900ppm chlorine (<1500ppm total Br + Cl) and <1000ppm antimony compounds.
4. For packaging details, go to our website at http"//www.diodes.com/products/packages.html.

Marking Information

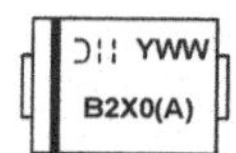

B2X0A = Product type marking code, ex: B220A (SMA package)
B2X0 = Product type marking code, ex: B230 (SMB package)
= Manufacturers' code marking
YWW = Date code marking
Y = Last digit of year (ex: 2 for 2002)
WW = Week code (01 to 53)

B220/A - B260/A
Document number: DS13004 Rev. 21 - 2

1 of 5
www.diodes.com

April 2013
© Diodes Incorporated

Figure A2.10. *Schottky diode: Diodes-Incorporated (Datasheet)*

B220/A - B260/A

Maximum Ratings (@T_A = +25°C, unless otherwise specified.)

Single phase, half wave, 60Hz, resistive or inductive load.
For capacitance load, derate current by 20%.

Characteristic	Symbol	B220/A	B230/A	B240/A	B250/A	B260/A	Unit
Peak Repetitive Reverse Voltage Working Peak Reverse Voltage DC Blocking Voltage	V_{RRM} V_{RWM} V_R	20	30	40	50	60	V
RMS Reverse Voltage	$V_{R(RMS)}$	14	21	28	35	42	V
Average Rectified Output Current @ T_L = +100°C	I_O	2.0					A
Non-Repetitive Peak Forward Surge Current, 8.3ms Single Half Sine-Wave Superimposed on Rated Load	I_{FSM}	50					A

Thermal Characteristics

Characteristic		Symbol	Value	Unit
Typical Thermal Resistance, Junction to Lead	SMA SMB	$R_{\theta JL}$	25 20	°C/W
Operating and Storage Temperature Range		T_J, T_{STG}	-65 to +150	°C

Electrical Characteristics (@T_A = +25°C, unless otherwise specified.)

Characteristic		Symbol	Min	Typ	Max	Unit	Test Condition
Forward Voltage Drop	B220/A, B230/A, B240/A B250/A, B260/A	V_F	—	—	0.50 0.70	V	I_F = 2.0A, T_A = +25°C
Leakage Current (Note 5)		I_R	— —	— —	0.5 20	mA	@ Rated V_R, T_A = +25°C @ Rated V_R, T_A = +100°C
Total Capacitance		C_T	—	—	200	pF	V_R = 4V, f = 1MHz

Note: 5. Short duration pulse test used to minimize self-heating effect.

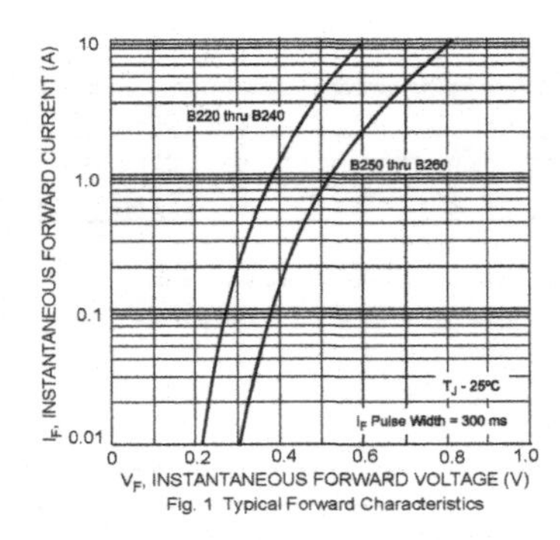

Fig. 1 Typical Forward Characteristics

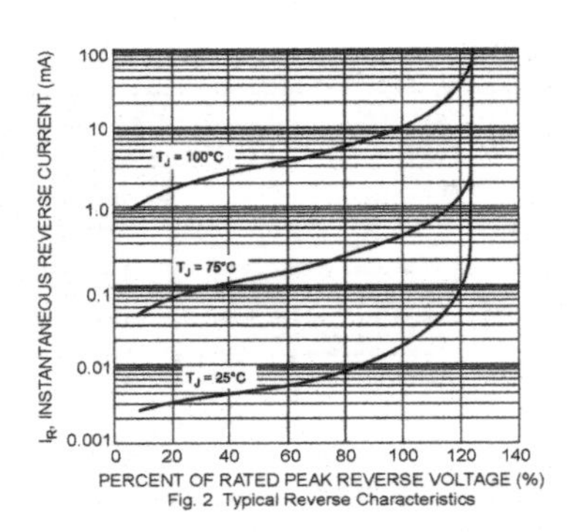

Fig. 2 Typical Reverse Characteristics

B220/A - B260/A
Document number: DS13004 Rev. 21 - 2

2 of 5
www.diodes.com

April 2013
© Diodes Incorporated

Figure A2.11. *Schottky diode: Diodes-Incorporated (Datasheet)*

DIODES INCORPORATED

B220/A - B260/A

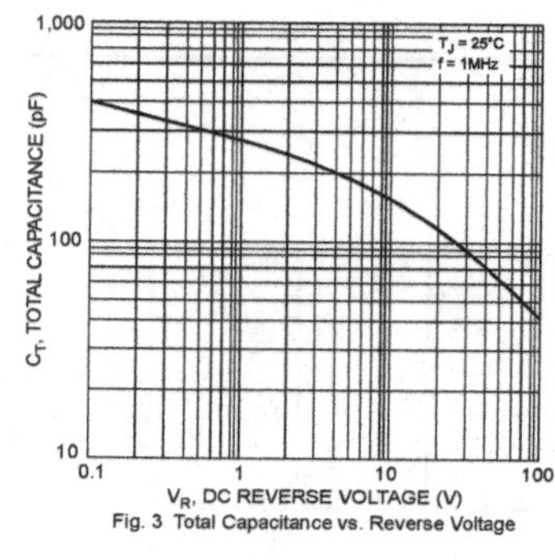

Fig. 3 Total Capacitance vs. Reverse Voltage

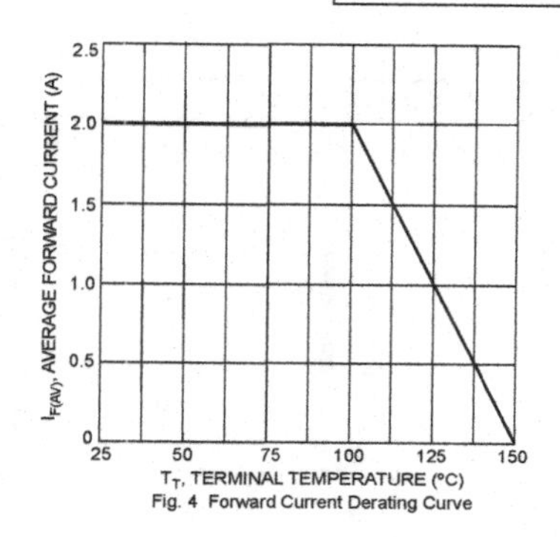

Fig. 4 Forward Current Derating Curve

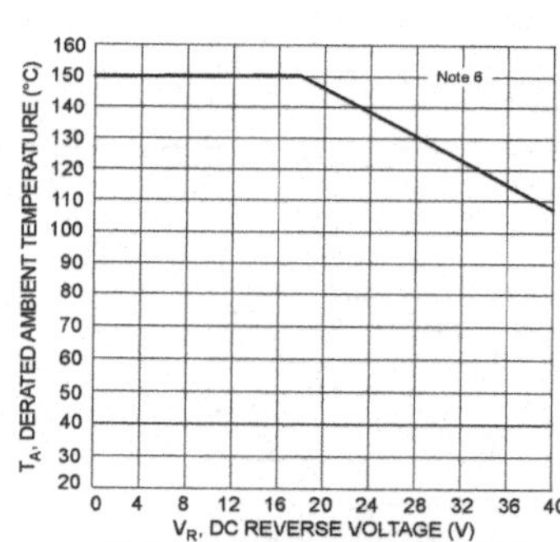

Fig. 5 Operating Temperature Derating (B240)

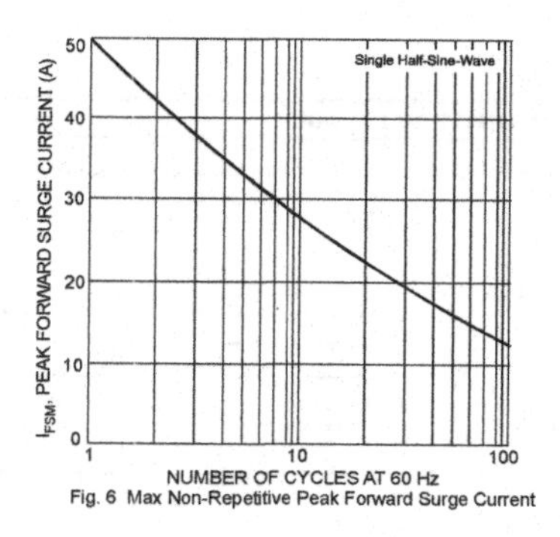

Fig. 6 Max Non-Repetitive Peak Forward Surge Current

6. Device mounted on FR-4 PC board with minimum recommended pad layout pattern as per http://www.diodes.com.

B220/A - B260/A
Document number: DS13004 Rev. 21 - 2

3 of 5
www.diodes.com

April 2013

Figure A2.12. *Schottky diode: Diodes-Incorporated (Datasheet)*

B220/A - B260/A

Package Outline Dimensions

Please see AP02002 at http://www.diodes.com/datasheets/ap02002.pdf for latest version.

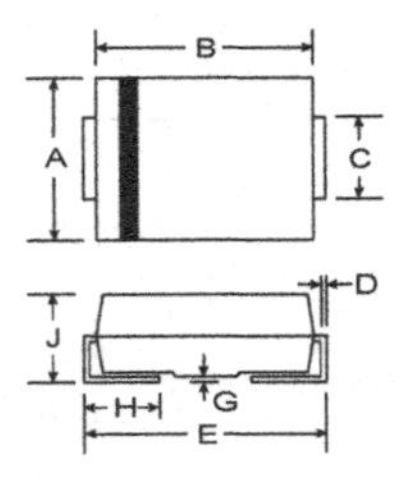

SMA		
Dim	Min	Max
A	2.29	2.92
B	4.00	4.60
C	1.27	1.63
D	0.15	0.31
E	4.80	5.59
G	0.05	0.20
H	0.76	1.52
J	2.01	2.30
All Dimensions in mm		

SMB		
Dim	Min	Max
A	3.30	3.94
B	4.06	4.57
C	1.96	2.21
D	0.15	0.31
E	5.00	5.59
G	0.05	0.20
H	0.76	1.52
J	2.00	2.50
All Dimensions in mm		

Suggested Pad Layout

Please see AP02001 at http://www.diodes.com/datasheets/ap02001.pdf for the latest version.

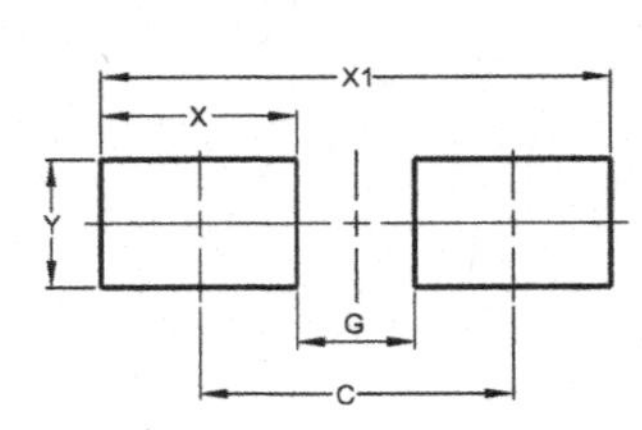

SMA	
Dimensions	Value (in mm)
C	4.00
G	1.50
X	2.50
X1	6.50
Y	1.70

SMB	
Dimensions	Value (in mm)
C	4.30
G	1.80
X	2.50
X1	6.80
Y	2.30

Figure A2.13. *Schottky diode: Diodes-Incorporated (Datasheet)*

A2.3. MOSFET power transistor driver

Only part of the technical documentation for the driver is shown below: the full 20-page document is available at www.microchip.com (taken from revision B of the document, Figures A2.14–A2.26).

MCP1415/16

Tiny 1.5A, High-Speed Power MOSFET Driver

Features

- High Peak Output Current: 1.5A (typical)
- Wide Input Supply Voltage Operating Range:
  - 4.5V to 18V
- Low Shoot-Through/Cross-Conduction Current in Output Stage
- High Capacitive Load Drive Capability:
  - 470 pF in 13 ns (typical)
  - 1000 pF in 20 ns (typical)
- Short Delay Times: 41 ns (t_{D1}), 48 ns (t_{D2}) (typical)
- Low Supply Current:
  - With Logic '1' Input - 0.65 mA (typical)
  - With Logic '0' Input - 0.1 mA (typical)
- Latch-Up Protected: will withstand 500 mA Reverse Current
- Logic Input will withstand Negative Swing up to 5V
- Space-saving 5L SOT-23 Package

Applications

- Switch Mode Power Supplies
- Pulse Transformer Drive
- Line Drivers
- Level Translator
- Motor and Solenoid Drive

General Description

MCP1415/16 devices are high-speed MOSFET drivers that are capable of providing 1.5A of peak current. The inverting or non-inverting single channel output is directly controlled from either TTL or CMOS (3V to 18V) logic. These devices also feature low shoot-through current, matched rise and fall time, and short propagation delays which make them ideal for high switching frequency applications.

MCP1415/16 devices operate from a single 4.5V to 18V power supply and can easily charge and discharge 1000 pF gate capacitance in under 20 ns (typical). They provide low enough impedances in both the on and off states to ensure that the intended state of the MOSFET will not be affected, even by large transients.

These devices are highly latch-up resistant under any condition within their power and voltage ratings. They are not subject to damage when noise spiking (up to 5V, of either polarity) occurs on the ground pin. They can accept, without damage or logic upset, up to 500 mA of reverse current being forced back into their outputs. All terminals are fully protected against Electrostatic Discharge (ESD) up to 2.0 kV (HBM) and 300V (MM).

Package Types:

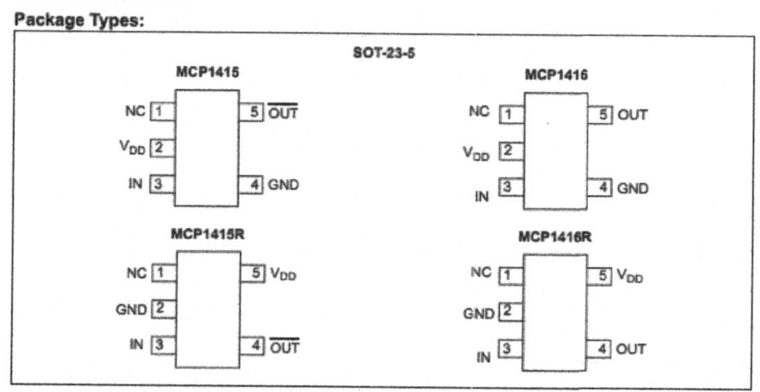

Figure A2.14. *MOSFET power transistor driver (Datasheet)*

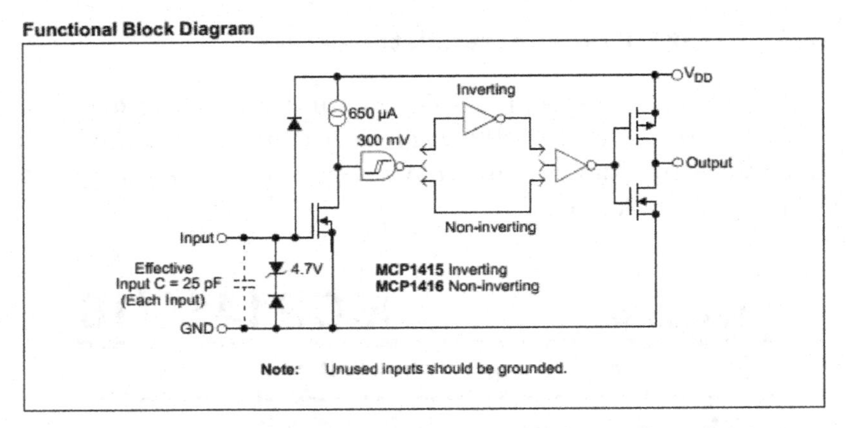

Figure A2.15. *MOSFET power transistor driver (Datasheet)*

MCP1415/16

1.0 ELECTRICAL CHARACTERISTICS

Absolute Maximum Ratings †

V_{DD}, Supply Voltage .... +20V
V_{IN}, Input Voltage .... (V_{DD} + 0.3V) to (GND - 5V)
Package Power Dissipation (T_A = 50°C)
5L SOT23 .... 0.39W
ESD Protection on all Pins .... 2.0 kV (HBM)
.... 400V (MM)

† **Notice:** Stresses above those listed under "Maximum Ratings" may cause permanent damage to the device. This is a stress rating only and functional operation of the device at those or any other conditions above those indicated in the operational sections of this specification is not intended. Exposure to maximum rating conditions for extended periods may affect device reliability.

DC CHARACTERISTICS

Electrical Specifications: Unless otherwise noted, T_A = +25°C, with 4.5V ≤ V_{DD} ≤ 18V

Parameters	Sym	Min	Typ	Max	Units	Conditions
Input						
Logic '1' High Input Voltage	V_{IH}	2.4	1.9	—	V	
Logic '0' Low Input Voltage	V_{IL}	—	1.6	0.8	V	
Input Current	I_{IN}	-1	—	+1	µA	0V ≤ V_{IN} ≤ V_{DD}
Input Voltage	V_{IN}	-5	—	V_{DD}+0.3	V	
Output						
High Output Voltage	V_{OH}	V_{DD} - 0.025	—	—	V	DC Test
Low Output Voltage	V_{OL}	—	—	0.025	V	DC Test
Output Resistance, High	R_{OH}	—	6	7.5	Ω	I_{OUT} = 10 mA, V_{DD} = 18V **(Note 2)**
Output Resistance, Low	R_{OL}	—	4	5.5	Ω	I_{OUT} = 10 mA, V_{DD} = 18V **(Note 2)**
Peak Output Current	I_{PK}	—	1.5	—	A	V_{DD} = 18V **(Note 2)**
Latch-Up Protection Withstand Reverse Current	I_{REV}	0.5	—	—	A	Duty cycle ≤ 2%, t ≤ 300 µs **(Note 2)**
Switching Time (Note 1)						
Rise Time	t_R	—	20	25	ns	Figure 4-1, Figure 4-2 C_L = 1000 pF **(Note 2)**
Fall Time	t_F	—	20	25	ns	Figure 4-1, Figure 4-2 C_L = 1000 pF **(Note 2)**
Delay Time	t_{D1}	—	41	50	ns	Figure 4-1, Figure 4-2 **(Note 2)**
Delay Time	t_{D2}	—	48	55	ns	Figure 4-1, Figure 4-2 **(Note 2)**
Power Supply						
Supply Voltage	V_{DD}	4.5	—	18	V	
Power Supply Current	I_S	—	0.65	1.1	mA	V_{IN} = 3V
	I_S	—	0.1	0.15	mA	V_{IN} = 0V

Note 1: Switching times ensured by design.
2: Tested during characterization, not production tested.

Figure A2.16. *MOSFET power transistor driver (Datasheet)*

MCP1415/16

DC CHARACTERISTICS (OVER OPERATING TEMPERATURE RANGE)

Electrical Specifications: Unless otherwise indicated, over operating range with 4.5V ≤ V_{DD} ≤ 18V.						
Parameters	**Sym**	**Min**	**Typ**	**Max**	**Units**	**Conditions**
Input						
Logic '1', High Input Voltage	V_{IH}	2.4	—	—	V	
Logic '0', Low Input Voltage	V_{IL}	—	—	0.8	V	
Input Current	I_{IN}	-10	—	+10	µA	0V ≤ V_{IN} ≤ V_{DD}
Input Voltage	V_{IN}	-5	—	V_{DD}+0.3	V	
Output						
High Output Voltage	V_{OH}	V_{DD} - 0.025	—	—	V	DC Test
Low Output Voltage	V_{OL}	—	—	0.025	V	DC Test
Output Resistance, High	R_{OH}	—	8.5	9.5	Ω	I_{OUT} = 10 mA, V_{DD} = 18V **(Note 2)**
Output Resistance, Low	R_{OL}	—	6	7	Ω	I_{OUT} = 10 mA, V_{DD} = 18V **(Note 2)**
Switching Time (Note 1)						
Rise Time	t_R	—	30	40	ns	Figure 4-1, Figure 4-2 C_L = 1000 pF **(Note 2)**
Fall Time	t_F	—	30	40	ns	Figure 4-1, Figure 4-2 C_L = 1000 pF **(Note 2)**
Delay Time	t_{D1}	—	45	55	ns	Figure 4-1, Figure 4-2 **(Note 2)**
Delay Time	t_{D2}	—	50	60		Figure 4-1, Figure 4-2 **(Note 2)**
Power Supply						
Supply Voltage	V_{DD}	4.5	—	18	V	
Power Supply Current	I_S	—	0.75	1.5	mA	V_{IN} = 3.0V
	I_S	—	0.15	0.25	mA	V_{IN} = 0V

Note 1: Switching times ensured by design.

2: Tested during characterization, not production tested.

TEMPERATURE CHARACTERISTICS

Electrical Specifications: Unless otherwise noted, all parameters apply with 4.5V ≤ V_{DD} ≤ 18V						
Parameter	**Sym**	**Min**	**Typ**	**Max**	**Units**	**Comments**
Temperature Ranges						
Specified Temperature Range	T_A	-40	—	+125	°C	
Maximum Junction Temperature	T_J	—	—	+150	°C	
Storage Temperature Range	T_A	-65	—	+150	°C	
Package Thermal Resistances						
Thermal Resistance, 5LD SOT23	θ_{JA}	—	256	—	°C/W	

Figure A2.17. *MOSFET power transistor driver (Datasheet)*

MCP1415/16

2.0 TYPICAL PERFORMANCE CURVES

Note:	The graphs and tables provided following this note are a statistical summary based on a limited number of samples and are provided for informational purposes only. The performance characteristics listed herein are not tested or guaranteed. In some graphs or tables, the data presented may be outside the specified operating range (e.g., outside specified power supply range) and therefore outside the warranted range.

Note: Unless otherwise indicated, T_A = +25°C with 4.5V ≤ V_{DD} ≤ = 18V.

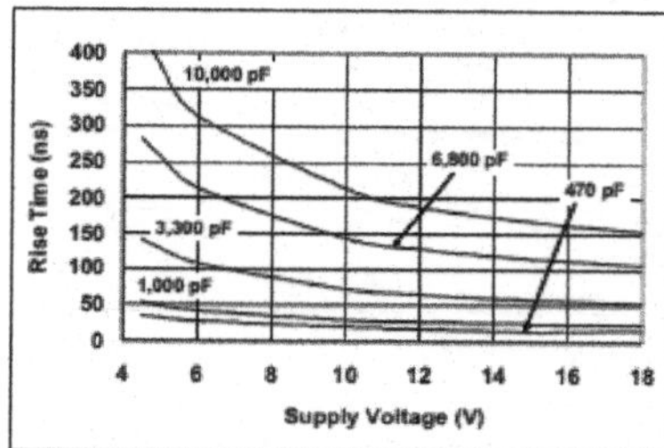

FIGURE 2-1: *Rise Time vs. Supply Voltage.*

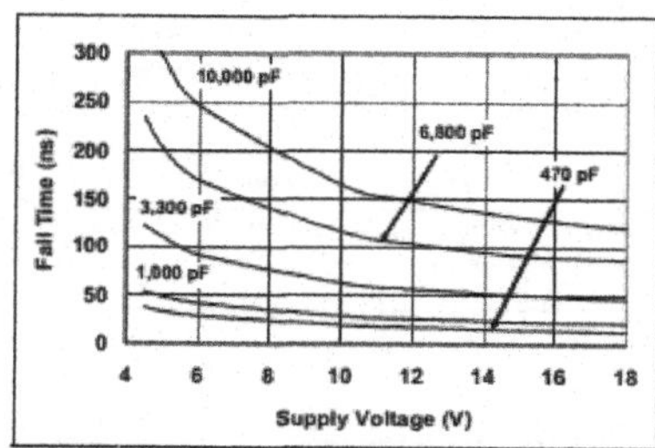

FIGURE 2-4: *Fall Time vs. Supply Voltage.*

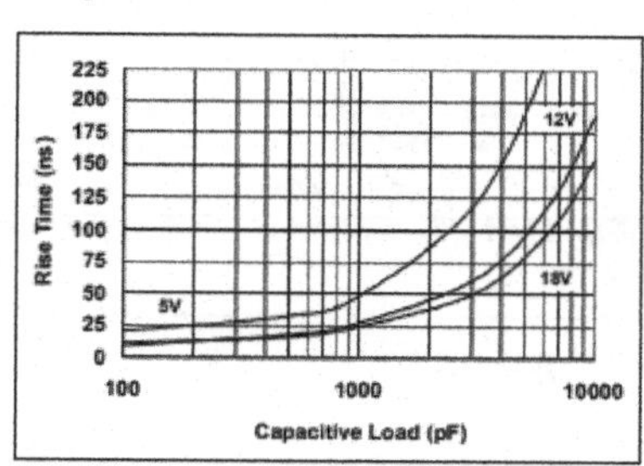

FIGURE 2-2: *Rise Time vs. Capacitive Load.*

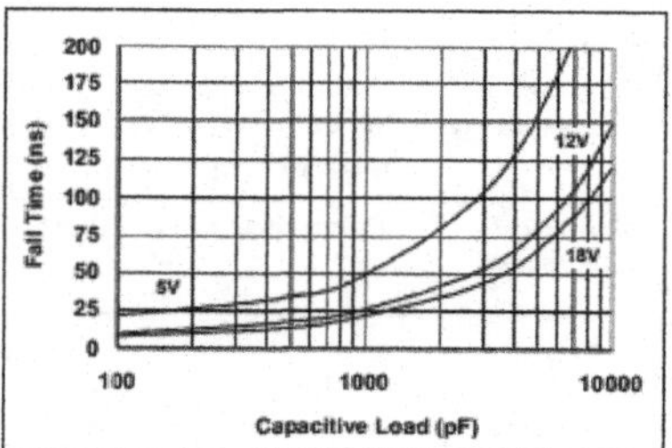

FIGURE 2-5: *Fall Time vs. Capacitive Load.*

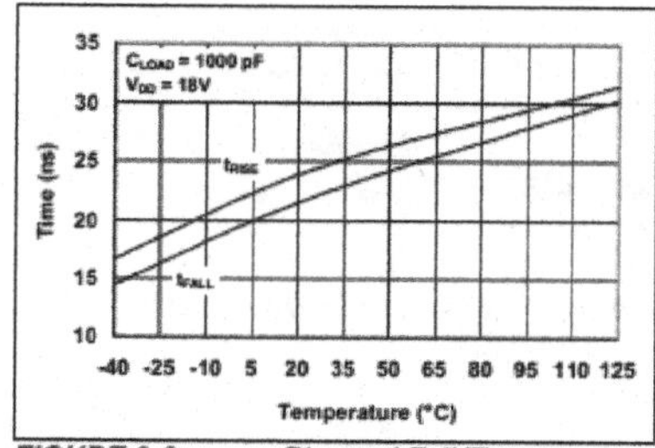

FIGURE 2-3: *Rise and Fall Times vs. Temperature.*

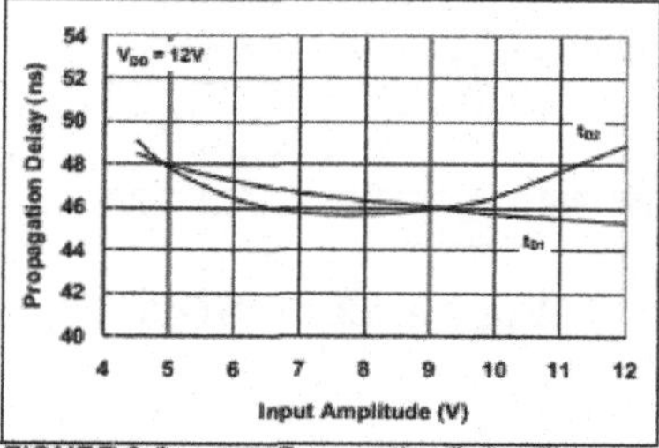

FIGURE 2-6: *Propagation Delay Time vs. Input Amplitude.*

Figure A2.18. *MOSFET power transistor driver (Datasheet)*

MCP1415/16

Note: Unless otherwise indicated, $T_A = +25°C$ with $4.5V \leq V_{DD} \leq = 18V$.

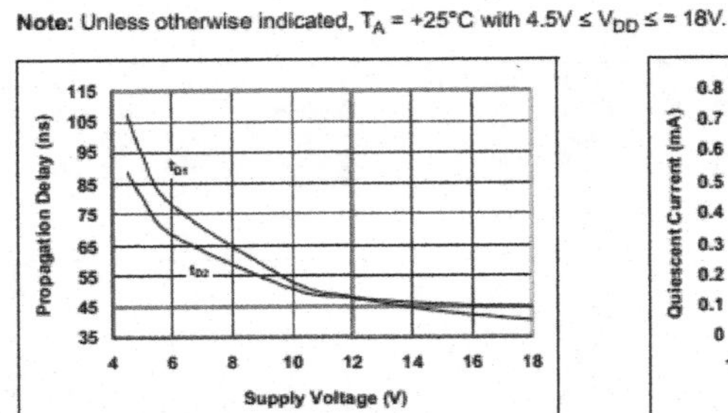

FIGURE 2-7: *Propagation Delay Time vs. Supply Voltage.*

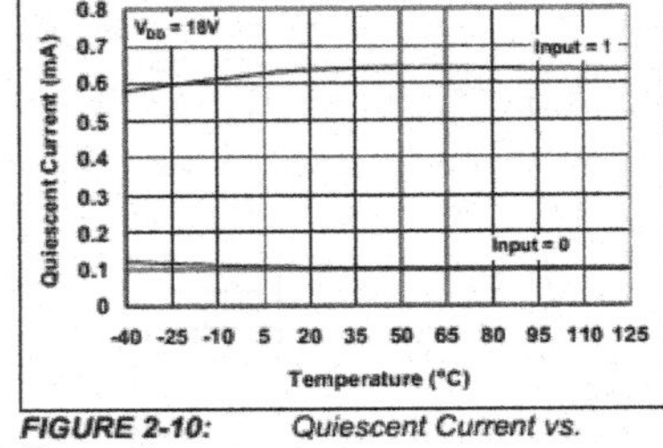

FIGURE 2-10: *Quiescent Current vs. Temperature.*

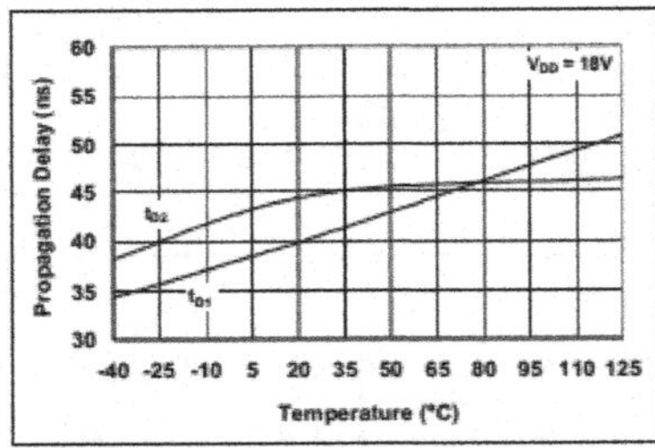

FIGURE 2-8: *Propagation Delay Time vs. Temperature.*

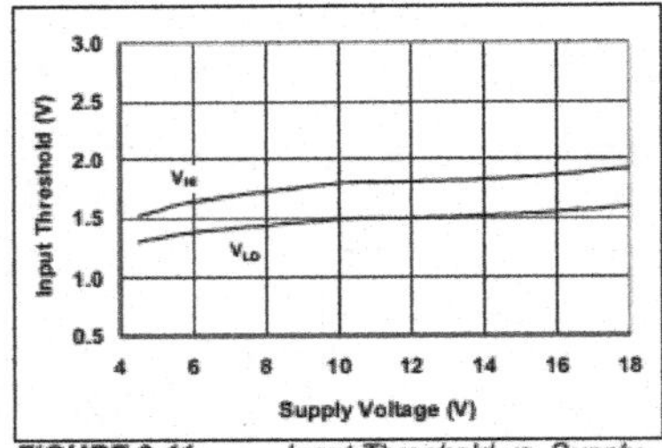

FIGURE 2-11: *Input Threshold vs. Supply Voltage.*

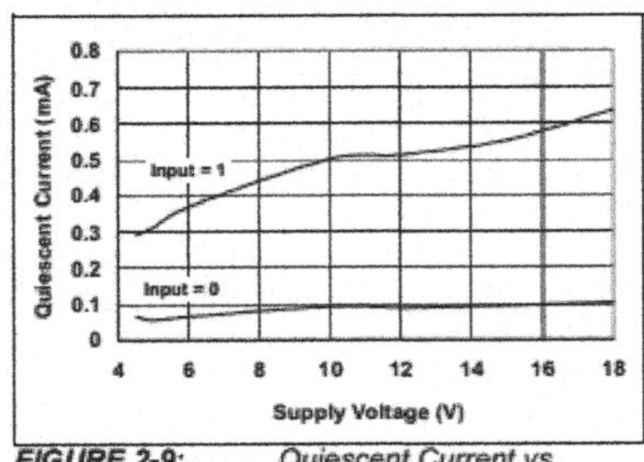

FIGURE 2-9: *Quiescent Current vs. Supply Voltage.*

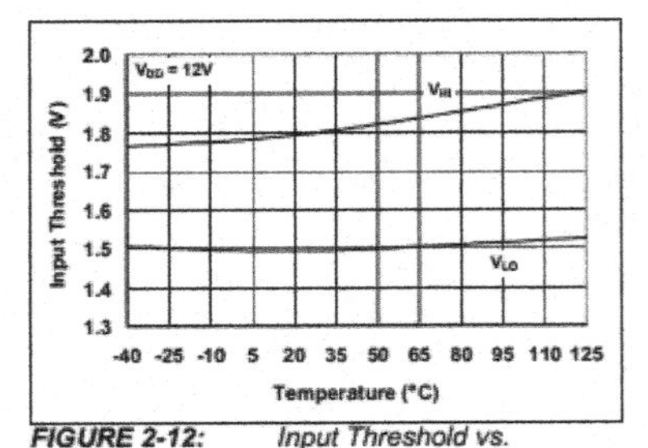

FIGURE 2-12: *Input Threshold vs. Temperature.*

Figure A2.19. *MOSFET power transistor driver (Datasheet)*

MCP1415/16

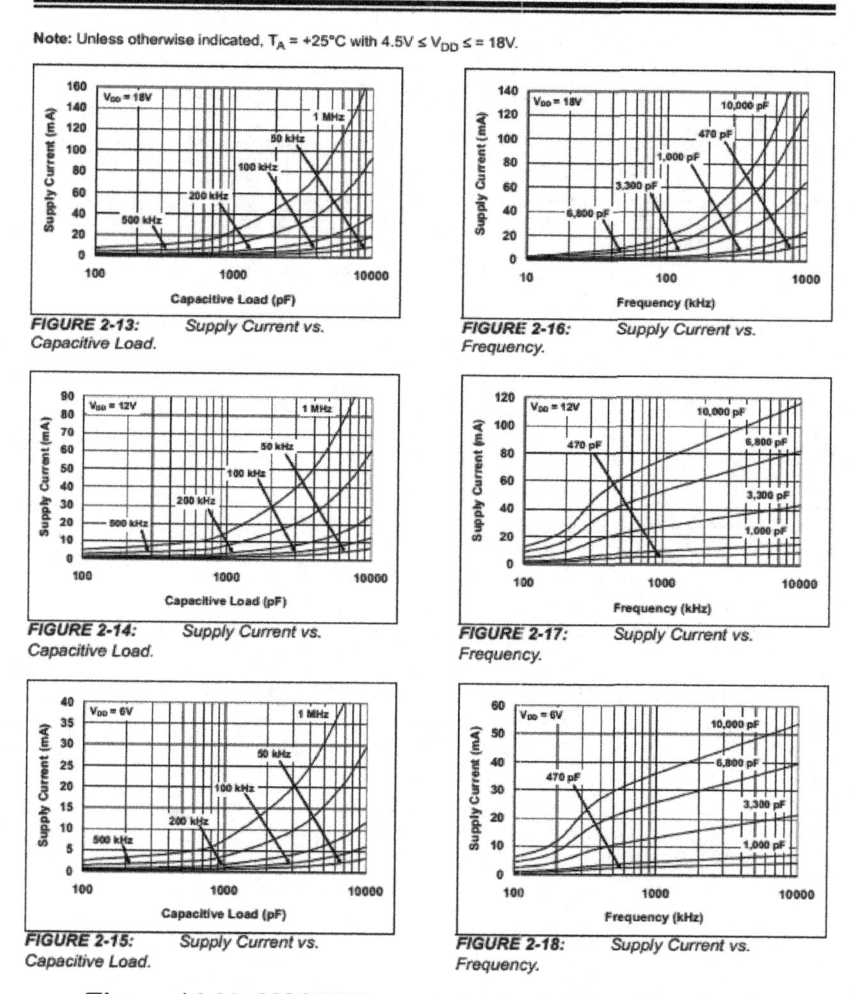

FIGURE 2-13: *Supply Current vs. Capacitive Load.*

FIGURE 2-14: *Supply Current vs. Capacitive Load.*

FIGURE 2-15: *Supply Current vs. Capacitive Load.*

FIGURE 2-16: *Supply Current vs. Frequency.*

FIGURE 2-17: *Supply Current vs. Frequency.*

FIGURE 2-18: *Supply Current vs. Frequency.*

Figure A2.20. *MOSFET power transistor driver (Datasheet)*

MCP1415/16

Note: Unless otherwise indicated, $T_A = +25°C$ with $4.5V \leq V_{DD} \leq = 18V$.

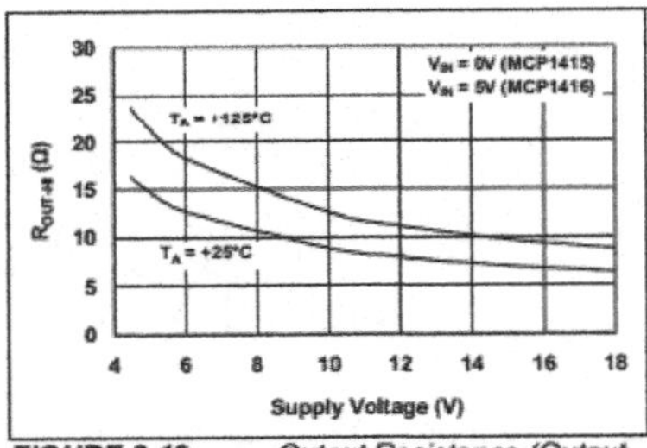

FIGURE 2-19: Output Resistance (Output High) vs. Supply Voltage.

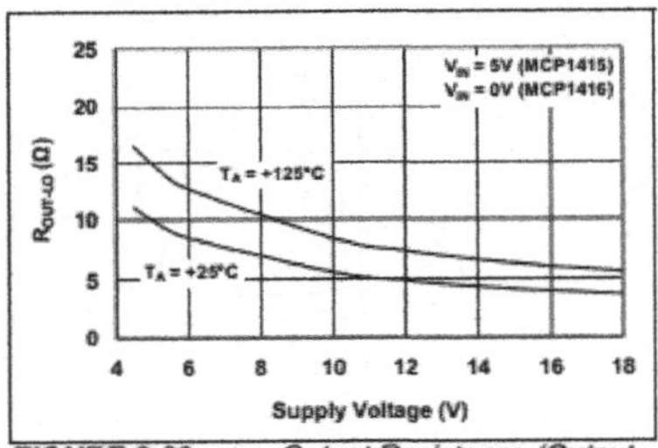

FIGURE 2-20: Output Resistance (Output Low) vs. Supply Voltage.

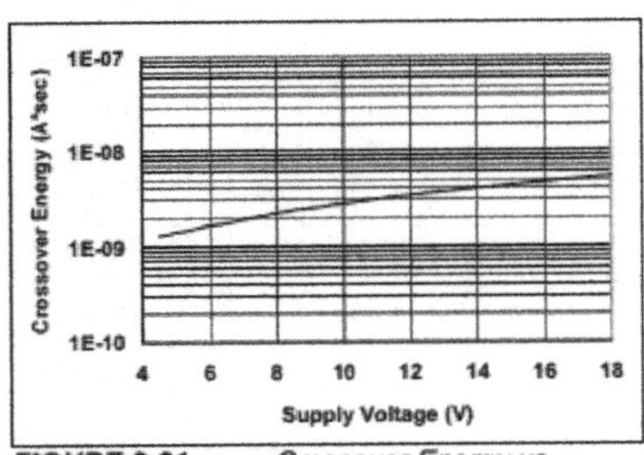

FIGURE 2-21: Crossover Energy vs. Supply Voltage.

Figure A2.21. *MOSFET power transistor driver (Datasheet)*

MCP1415/16

3.0 PIN DESCRIPTIONS

The descriptions of the pins are listed in Table 3-1.

TABLE 3-1: PIN FUNCTION TABLE

SOT-23-5	Symbol		Description
Pin	MCP1415/6	MCP1415R/6R	
1	NC	NC	No Connection
2	V_{DD}	GND	Supply Input
3	IN	IN	Control Input
4	GND	OUT	Ground
5	OUT	V_{DD}	Output

3.1 Supply Input (V_{DD})

V_{DD} is the bias supply input for the MOSFET driver and has a voltage range of 4.5V to 18V. This input must be decoupled to ground with a local capacitor. This bypass capacitor provides a localized low impedance path for the peak currents that are to be provided to the load.

3.2 Control Input (IN)

The MOSFET driver input is a high impedance, TTL/CMOS compatible input. The input also has hysteresis between the high and low input levels, allowing them to be driven from a slow rising and falling signals, and to provide noise immunity.

3.3 Ground (GND)

Ground is the device return pin. The ground pin should have a low impedance connection to the bias supply source return. High peak currents will flow out the ground pin when the capacitive load is being discharged.

3.4 Output (OUT)

The output is a CMOS push-pull output that is capable of sourcing and sinking 1.5A of peak current (V_{DD} = 18V). The low output impedance ensures the gate of the external MOSFET will stay in the intended state even during large transients. This output also has a reverse current latch-up rating of 500 mA.

Figure A2.22. *MOSFET power transistor driver (Datasheet)*

MCP1415/16

4.0 APPLICATION INFORMATION

4.1 General Information

MOSFET drivers are high-speed, high current devices which are intended to source/sink high peak currents to charge/discharge the gate capacitance of external MOSFETs or IGBTs. In high frequency switching power supplies, the PWM controller may not have the drive capability to directly drive the power MOSFET. A MOSFET driver like the MCP1415/16 family can be used to provide additional source/sink current capability.

4.2 MOSFET Driver Timing

The ability of a MOSFET driver to transition from a fully-off state to a fully-on state are characterized by the drivers rise time (t_R), fall time (t_F), and propagation delays (t_{D1} and t_{D2}). The MCP1415/16 family of drivers can typically charge and discharge a 1000 pF load capacitance in 20 ns along with a typical turn on (t_{D1}) propagation delay of 41 ns. Figure 4-1 and Figure 4-2 show the test circuit and timing waveform used to verify the MCP1415/16 timing.

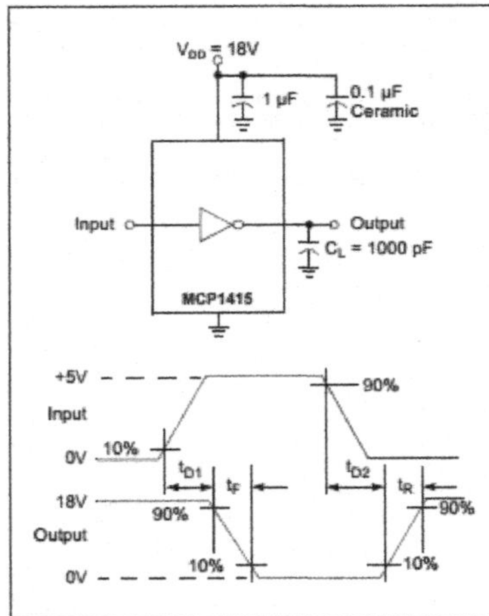

FIGURE 4-1: Inverting Driver Timing Waveform.

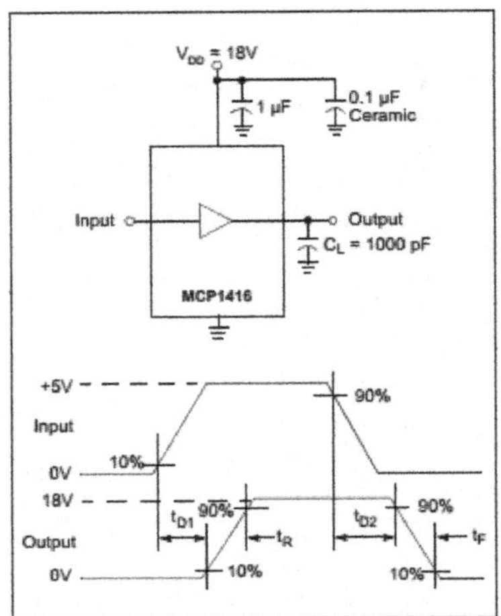

FIGURE 4-2: Non-Inverting Driver Timing Waveform.

4.3 Decoupling Capacitors

Careful layout and decoupling capacitors are required when using power MOSFET drivers. Large current are required to charge and discharge capacitive loads quickly. For example, approximately 720 mA are needed to charge a 1000 pF load with 18V in 25 ns.

To operate the MOSFET driver over a wide frequency range with low supply impedance, a ceramic and a low ESR film capacitor are recommended to be placed in parallel between the driver V_{DD} and GND. A 1.0 µF low ESR film capacitor and a 0.1 µF ceramic capacitor placed between pins 2 and 4 is required for reliable operation. These capacitors should be placed close to the driver to minimize circuit board parasitics and provide a local source for the required current.

Figure A2.23. *MOSFET power transistor driver (Datasheet)*

4.4 Power Dissipation

The total internal power dissipation in a MOSFET driver is the summation of three separate power dissipation elements.

EQUATION 4-1:

$$P_T = P_L + P_Q + P_{CC}$$

Where:

P_T	=	Total power dissipation
P_L	=	Load power dissipation
P_Q	=	Quiescent power dissipation
P_{CC}	=	Operating power dissipation

4.4.1 CAPACITIVE LOAD DISSIPATION

The power dissipation caused by a capacitive load is a direct function of the frequency, total capacitive load, and supply voltage. The power lost in the MOSFET driver for a complete charging and discharging cycle of a MOSFET is shown in Equation 4-2.

EQUATION 4-2:

$$P_L = f \times C_T \times V_{DD}^2$$

Where:

f	=	Switching frequency
C_T	=	Total load capacitance
V_{DD}	=	MOSFET driver supply voltage

4.4.2 QUIESCENT POWER DISSIPATION

The power dissipation associated with the quiescent current draw depends upon the state of the input pin. The MCP1415/16 devices have a quiescent current draw when the input is high of 0.65 mA (typical) and 0.1 mA (typical) when the input is low. The quiescent power dissipation is shown in Equation 4-3.

EQUATION 4-3:

$$P_Q = (I_{QH} \times D + I_{QL} \times (1 - D)) \times V_{DD}$$

Where:

I_{QH}	=	Quiescent current in the high state
D	=	Duty cycle
I_{QL}	=	Quiescent current in the low state
V_{DD}	=	MOSFET driver supply voltage

4.4.3 OPERATING POWER DISSIPATION

The operating power dissipation occurs each time the MOSFET driver output transitions because for a very short period of time both MOSFETs in the output stage are on simultaneously. This cross-conduction current leads to a power dissipation describe in Equation 4-4.

EQUATION 4-4:

$$P_{CC} = CC \times f \times V_{DD}$$

Where:

CC	=	Cross-conduction constant (A*sec)
f	=	Switching frequency
V_{DD}	=	MOSFET driver supply voltage

4.5 PCB Layout Considerations

Proper PCB layout is important in high current, fast switching circuits to provide proper device operation and robustness of design. Improper component placement may cause errant switching, excessive voltage ringing, or circuit latch-up. PCB trace loop area and inductance must be minimized. This is accomplished by placing the MOSFET driver directly at the load and placing the bypass capacitor directly at the MOSFET driver (Figure 4-3). Locating ground planes or ground return traces directly beneath the driver output signal also reduces trace inductance. A ground plane will also help as a radiated noise shield as well as providing some heat sinking for power dissipated within the device (Figure 4-4).

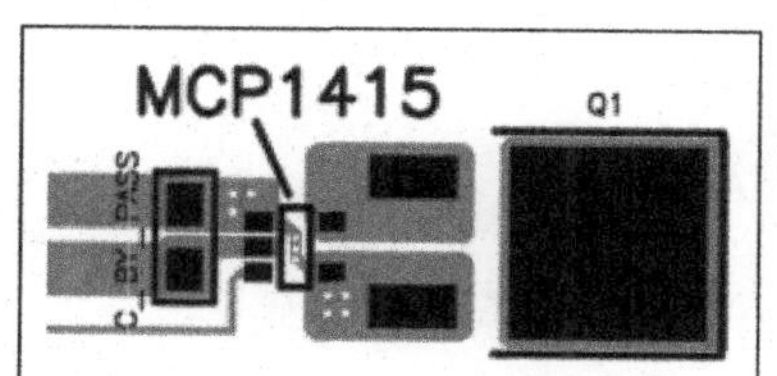

FIGURE 4-3: Recommended PCB Layout (TOP).

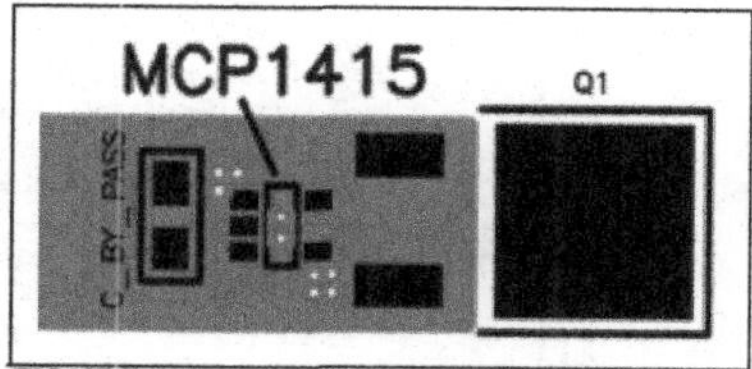

FIGURE 4-4: Recommended PCB Layout (BOTTOM).

Figure A2.24. *MOSFET power transistor driver (Datasheet)*

MCP1415/16

5.0 PACKAGING INFORMATION

5.1 Package Marking Information

5-Lead SOT-23

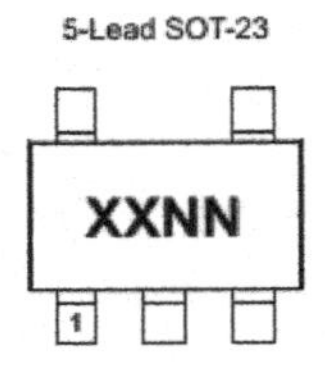

Standard Markings for SOT-23	
Part Number	**Code**
MCP1415T-E/OT	FYNN
MCP1416T-E/OT	FZNN
MCP1415RT-E/OT	F7NN
MCP1416RT-E/OT	F8NN

Example

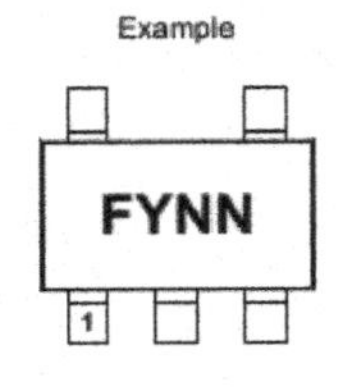

Legend:
- XX...X Customer-specific information
- Y Year code (last digit of calendar year)
- YY Year code (last 2 digits of calendar year)
- WW Week code (week of January 1 is week '01')
- NNN Alphanumeric traceability code
- (e3) Pb-free JEDEC designator for Matte Tin (Sn)
- * This package is Pb-free. The Pb-free JEDEC designator (e3) can be found on the outer packaging for this package.

Note: In the event the full Microchip part number cannot be marked on one line, it will be carried over to the next line, thus limiting the number of available characters for customer-specific information.

Figure A2.25. *MOSFET power transistor driver (Datasheet)*

MCP1415/16

5-Lead Plastic Small Outline Transistor (OT) [SOT-23]

Note: For the most current package drawings, please see the Microchip Packaging Specification located at http://www.microchip.com/packaging

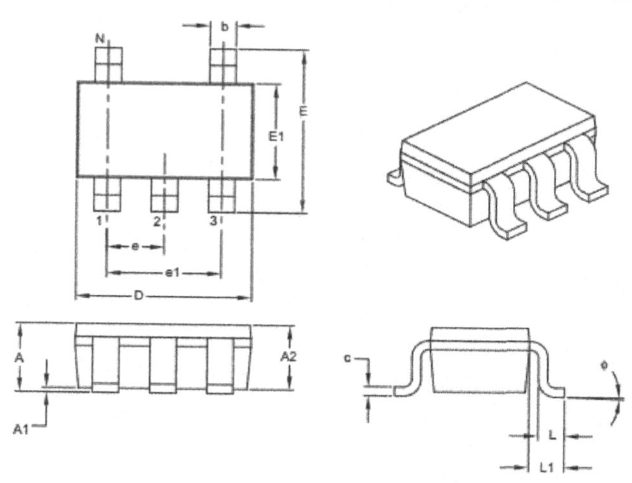

Units		MILLIMETERS		
Dimension Limits		MIN	NOM	MAX
Number of Pins	N		5	
Lead Pitch	e		0.95 BSC	
Outside Lead Pitch	e1		1.90 BSC	
Overall Height	A	0.90	–	1.45
Molded Package Thickness	A2	0.89	–	1.30
Standoff	A1	0.00	–	0.15
Overall Width	E	2.20	–	3.20
Molded Package Width	E1	1.30	–	1.80
Overall Length	D	2.70	–	3.10
Foot Length	L	0.10	–	0.60
Footprint	L1	0.35	–	0.80
Foot Angle	φ	0°	–	30°
Lead Thickness	c	0.08	–	0.26
Lead Width	b	0.20	–	0.51

Notes:

1. Dimensions D and E1 do not include mold flash or protrusions. Mold flash or protrusions shall not exceed 0.127 mm per side.
2. Dimensioning and tolerancing per ASME Y14.5M.

   BSC: Basic Dimension. Theoretically exact value shown without tolerances.

Microchip Technology Drawing C04-091B

Figure A2.26. *MOSFET power transistor driver (Datasheet)*

Bibliography

[APP 02] APPEL W., *Mathématiques pour la physique et les physiciens*, H & K, Paris, 2002.

[BAS 09] BASDEVANT J.-L., *Les mathématiques de la physique quantique*, Vuibert, Paris, 2009.

[BEC 00] BECH M.M., Analysis of random pulse-width modulation techniques for power electronic converters, PhD Thesis, Aalborg University, 2000.

[BEN 33] BENNETT W.R., "New results in the calculation of modulation productions", *Bell Labs Technical Journal*, vol. 12, no. 4, pp. 238–243, 1933.

[BIE 08] BIERHOFF M., FUCHS F.W., "DC link harmonics of three phase voltage source converters influenced by the pulsewidth modulation strategy – an analysis", *IEEE Transactions on Industrial Electronics*, vol. 55, no. 5, pp. 2085–2092, May 2008.

[BLA 53] BLACK H.S., *Modulation Theories*, Van Nostrand, New York, 1953.

[BOW 08] BOWICK C., BLYLER J., AJLUNI C., *RF Circuit Design*, Newnes/Elsevier, Boston, Amsterdam, 2008.

[BRE 05] BRÉHAUT S., Modélisation et optimisation des performances CEM d'un convertisseur AC/DC d'une puissance de 600 W, Phd Thesis, Université François Rabelais, Tours, 2005.

[BUH 91] BÜHLER H., *Convertisseurs statiques*, Presses Polytechniques et Universitaires Romandes, Lausanne, 1991.

[CHA 05] CHAROY A., *Compatibilité électromagnétique*, Dunod, Paris, 2005.

[CHE 09] CHENAND L., PENG F.Z., "Closed-loop gate drive for high power IGBTs", *Proceedings of IEEE Applied Power Electronics Conference (APEC)*, pp. 1331–1337, 2009.

[CHE 99] CHERON Y., *La commutation douce*, Tec & Doc, Paris, 1999.

[COH 00] COHEN DE LARA M., D'ANDRÉA-NOVEL B., *Cours d'automatique – commande linéaire des systèmes dynamiques*, Presses des Mines, Paris, 2000.

[COC 02] COCQUERELLE J.-L., PASQUIER C., *Rayonnement électromagnétique des convertisseurs à découpage*, EDP Sciences, Les Ulis, 2002.

[COR 13] CORNELL D., Aluminum electrolytic capacitor application guide, 2013. Available at: www.cde.com/fliptest/alum/alum.html.

[COS 05] COSTA F., MAGNON D., "Graphical analysis of the spectra of EMI sources in power electronics", *IEEE Transactions on Power Electronics*, vol. 20, no. 6, pp. 1491–1498, November 2005.

[COS 13] COSTA F., GAUTIER C., LABOURÉ E. *et al.*, *La compatibilité électromagnétique en électronique de puissance, Principes et cas d'études*, Hermès-Lavoisier, Paris, 2013.

[DEG 01] DEGRANGE B., *Introduction à la physique quantique*, Presses des Mines, Paris, 2001.

[FER 02] FERRIEUX J.-P., FOREST F., *Alimentations à découpages et convertisseurs à résonance*, 3rd ed., Dunod, Paris, 2002.

[FEY 99] FEYNMAN R., LEIGHTON R.B., SANDS M., *Cours de physique de Feynman, Electromagnétisme*, Dunod, Paris, vol. 1–2, 1999.

[FOC 98] FOCH H., FOREST F., MEYNARD T., "Onduleurs de tension – structures, principes, applications", *Techniques de l'Ingénieur*, Traité Génie Electrique, Article D3176, 1998.

[FOC 11] FOCH H., CHÉRON Y., "Convertisseur de type forward – dimensionnement", *Techniques de l'ingénieur*, Traité Génie Electrique, Article D3167, 2011.

[FRI 94] FRICKEY D.A., "Conversions between S, Z, Y, h, ABCD, and T parameters which are valid for complex source and load impedances", *IEEE Transactions on Microwave Theory and Techniques*, vol. 42, no. 2, 1994.

[GHA 03] GHAUSI M., LAKER K., *Modern Filter Design: Active RC and Switched Capacitor*, Noble Publishing, Atlanta, 2003.

[GIB 07] GIBSON W.C., *The Method of Moments in Electromagnetics*, Chapman & Hall/CRC, Boca Raton, 2007.

[HAV 98] HAVA A.M., KERKMAN R.J., LIPO T.A., "A high performance generalized discontinuous PWM algorithm", *IEEE Transactions on Industry Applications*, vol. 34, no. 5, pp. 1059–1071, 1998.

[HAV 99] HAVA A.M., LIPO T.A., KERKMAN R.J., "Simple analytical and graphical methods for carrier-based PWM-VSI drives", *IEEE Transactions on Power Electronics*, vol. 14, no. 1, pp. 49–61, 1999.

[HOB 05] HOBRAICHE J., Comparaison des stratégies de modulation à largeur d'impulsions triphasées – application à l'alterno-démarreur, PhD Thesis, UTC, Compiègne, 2005.

[HOL 83] HOLTZ J., STADTFELD S., "A predictive controller for a stator current vector of AC-machines fed from a switched voltage source", *International Power Electronics Conference IPEC*, vol. 2, pp. 1665–1675, Tokyo, 1983.

[IEE 12] IEEE Electromagnetic Compatibility Standards Collection: VuSpec^{TM}, CD-ROM, IEEE Standards Association, 2012.

[KEM 12] KEMET, Electrolytic Capacitors, Documentation technique, FF3304 06/09, 2012. Available at: www.kemet.com.

[KOL 91] KOLAR J.W., ERLT H., ZACH F.C., "Influence of the modulation method on the conduction and switching losses of a PWM converter system", *IEEE Transactions on Industry Applications*, vol. 27, no. 6, pp. 399–403, 1991.

[KOL 93] KOLAR J.W., Brevet intitulé "Vorrichtung und Verfahren zur Umformung von Drehstrom in Gleichstrom", Inventor: Déposant: IXYS Semiconductors GmdH, date de priorité: 23 December 1993, date of publication: 28 June 1995.

[LAN 09] LANFRANCHI V., PATIN N., DÉPERNET D., "MLI précalculées et optimisées", in MONMASSON E. (ed.), *Commande Rapprochée de Convertisseurs*, Chapter 4, Hermès-Lavoisier, Paris, 2009.

[LEF 02] LEFEBVRE S., MULTON B., "Commande des semi-conducteurs de puissance: principes", *Techniques de l'Ingénieur*, D-3231, 2002.

[LES 97] LESBROUSSARD C., Etude d'une stratégie de modulation de largeur d'impulsions pour un onduleur de tension triphasé à deux ou trois niveaux: la Modulation Delta Sigma Vectorielle, PhD Thesis, UTC, 1997.

[LIN 10] LINDER A., KANCHAN R., KENNEL R., *et al.*, *Model Based Predictive Control of Electrical Drives*, Cuvillier Verlag, Gottingen, 2010.

[LUM 00] LUMBROSO H., *Ondes électromagnétiques dans le vide et les conducteurs, 70 problèmes résolus*, 2nd ed., Dunod, Paris, 2000.

[MAT 09] MATHIEU H., FANET H., *Physique des semiconducteurs et des composants*, 6th ed., Dunod, Paris, 2009.

[MEY 93] MEYNARD T., FOCH H., "Imbricated cells multilevel voltage-source inverters for high-voltage applications", *EPE Journal*, vol. 3, June 1993.

[MIC 05] MICROCHIP, "Sinusoidal Control of PMSM Motors with dsPIC30F3010", AN1017, 2005. Available at: ww1.microchip.com/downloads/en/AppNotes/01017A.pdf.

[MID 77] MIDDLEBROOK R.D., CÙK S., "A general unified approach to modeling switching converter power stages", *International Journal of Electronics*, vol. 42, no. 6, pp. 521–550, 1977.

[MOH 95] MOHAN N., UNDELAND T.M., ROBBINS W.P., *Power Electronics – Converters, Applications and Design*, 2nd ed., Wiley, New York, 1995.

[MON 97] MONMASSON E., FAUCHER J., "Projet pédagogique autour de la MLI vectorielle", *Revue 3EI*, no. 8, pp. 22–36, 1997.

[MON 09] MONMASSON E. (ed.), *Commande rapprochée de convertisseurs statiques 1, Modulation de largeur d'impulsion*, Hermès-Lavoisier, Paris, 2009.

[MON 11] MONMASSON E. (ed.), *Power Electronic Converters: PWM Strategies and Current Control Techniques*, ISTE, London and John Wiley & Sons, New York, 2011.

[MOR 07] MOREL F., Commandes directes appliquées à une machine synchrone à aimants permanents alimentée par un onduleur triphasé à deux niveaux ou par un convertisseur matriciel triphasé, PhD Thesis, INSA de Lyon, 2007.

[MOY 98] MOYNIHAN J.F., EGAN M.G., MURPHY J.M.D., "Theoretical spectra of space vector modulated waveforms", *IEE Proceedings – Electric Power Applications*, vol. 145, pp. 17–24, 1998.

[MUK 10] MUKHTAR A., *High Performance AC Drives*, Springer, Berlin, 2010.

[NAR 06] NARAYANAN G., KRISHNAMURTHY H.K., ZHAO D. *et al.*, "Advanced bus-clamping PWM techniques based on space vector approach", *IEEE Transactions on Power Electronics*, vol. 21, no. 4, pp. 974–984, 2006.

[NAR 08] NARAYANAN G., RANGANATHAN V.T., ZHAO D. *et al.*, "Space vector based hybrid PWM techniques for reduced current ripple", *IEEE Transactions on Industrial Electronics*, vol. 55, no. 4, pp. 1614–1627, 2008.

[NGU 11a] NGUYEN T.D., Etude de stratégies de modulation pour onduleur triphasé dédiées à la réduction des perturbations du bus continu en environnement embarqué, PhD Thesis, UTC, Compiègne, 2011.

[NGU 11b] NGUYEN T.D., HOBRAICHE J., PATIN N. *et al.*, "A direct digital technique implementation of general discontinuous pulse width modulation strategy", *IEEE Transactions on Industrial Electronics*, vol. 58, no. 9, pp. 4445–4454, September 2011.

[OSW 11] OSWALD N., STARK B., HOLLIDAY D. *et al.*, "Analysis of shaped pulse transitions in power electronic switching waveforms for reduced EMI generation", *IEEE Transactions on Industry Applications*, vol. 47, no. 5, pp. 2154–2165, October-November 2011.

[PAT 15a] PATIN N., *Power Electronics Applied to Industrial Systems and Transports – Volume 1*, ISTE Press, London and Elsevier, Oxford, 2015.

[PAT 15b] PATIN N., *Power Electronics Applied to Industrial Systems and Transports – Volume 2*, ISTE Press, London and Elsevier, Oxford, 2015.

[PAT 15c] PATIN N., *Power Electronics Applied to Industrial Systems and Transports – Volume 4*, ISTE Press, London and Elsevier, Oxford, 2015.

[REB 98] REBY F., BAUSIERE R., SOHIER B. *et al.*, "Reduction of radiated and conducted emissions in power electronic circuits by the continuous derivative control method (CDCM)", *Proceedings of 7th International Conference on Power Electronics and Variable Speed Drives*, pp. 158–162, 1998.

[REV 03] REVOL B., Modélisation et optimisation des performances CEM d'une association variateur de vitesse machine asynchrone, PhD Thesis, Université Joseph Fourier, 2003.

[ROM 86] ROMBAULT C., SÉGUIER G., BAUSIÈRE R., *L'électronique de puissance – Volume 2, La conversion AC-AC*, Tec & Doc, Paris, 1986.

[ROU 04a] ROUDET J., CLAVEL E., GUICHON J.M. *et al.*, "Modélisation PEEC des connexions dans les convertisseurs de puissance", *Techniques de l'Ingénieur*, D-3071, 2004.

[ROU 04b] ROUDET J., CLAVEL E., GUICHON J.M. *et al.*, "Application de la méthode PEEC au cablage d'un onduleur triphasé", *Techniques de l'Ingénieur*, D-3072, 2004.

[ROU 04c] ROUSSEL J.-M., *Problèmes d'électronique de puissance*, Dunod, Paris, 2004.

[SCH 99] SCHELLMANNS A., Circuits équivalents pour transformateurs multienroulements, Application à la CEM conduite d'un convertisseur, PhD Thesis, l'INPG, July 1999.

[SCH 01] SCHNEIDER E., DELABALLE J., "La CEM: la compatibilité électromagnétique", *Cahier Technique*, no. 149, 2001.

[SEG 11] SÉGUIER G., DELARUE P., LABRIQUE F., *Electronique de puissance*, Dunod, Paris, 2011.

[SHU 11] SHUKLA A., GHOSH A., JOSHI A., "Natural balancing of flying capacitor voltages in multicell inverter under PD carrier-based PWM", *IEEE Transactions on Power Electronics*, vol. 56, no. 6, pp. 1682–1693, June 2011.

[VAS 12] VASCAS F., IANNELI L. (eds), *Dynamics and Control of Switched Electronic Systems*, Springer, Berlin, 2012.

[VEN 07] VENET P., Amélioration de la sûreté de fonctionnement des dispositifs de stockage d'énergie, mémoire d'HDR, University Claude Bernard – Lyon 1, 2007

[VIS 07] VISSER J.H., Active converter based on the vienna rectifier topology interfacing a three-phase generator to a DC-Bus, PhD Thesis, University of Pretoria, Afrique du Sud, 2007.

[VIS 12] VISSER H.J., *Antenna Theory and Applications*, Wiley, New York, 2012.

[VOG 11] VOGELSBERGER M.A., WIESINGER T., ERTL H., "Life-cycle monitoring and voltage-managing unit for DC-link electrolytic capacitors in PWM converters", *IEEE Transactions on Power Electronics*, vol. 26, no. 2, pp. 493–503, February 2011.

[WEE 06] WEENS Y., Modélisation des câbles d'énergie soumis aux contraintes générées par les convertisseurs électroniques de puissance, PhD Thesis, USTL, Lille, 2006.

[WHE 04] WHEELER P.W., CLARE J.C., EMPRINGHAM L. *et al.*, "Matrix converters", *IEEE Industry Applications Magazine*, vol. 10, no. 1, pp. 59–65, January-February 2004.

[WIL 99] WILLIAMS T., *Compatibilité électromagnétique – De la conception à l'homologation*, Publitronic, Paris, 1999.

[XAP 05] ALEXANDER M., Power Distribution System (PDS) Design: Using Bypass/Decoupling Capacitors, Application Note, XAPP623, available at: www.xilinx.com, February 2005.

[YUA 00] YUAN X., BARBI I., "Fundamentals of a new diode clamping multilevel inverter", *IEEE Transactions on Power Electronics*, vol. 15, no. 4, pp. 711–718, July 2000.

[ZHA 10] ZHAO D., HARI V.S.S.P.K., NARAYANAN G. *et al.*, "Space-vector-based hybrid pulsewidth modulation techniques for reduced harmonic distortion and switching loss", *IEEE Transactions on Power Electronics*, vol. 25, no. 3, pp. 760–774, 2010.

Index

C

chopper, 1, 2, 7, 9, 23, 66, 72
Clarke transformation, 114–118
closed-loop, 69, 72, 97–101, 108
converter
  boost, 8, 9, 13, 17, 39, 67
  buck, 1, 3–5, 8, 9, 19, 31, 61, 66
  coupled inductance (Flyback), 3, 36, 71, 72, 82, 83, 92
current
  branch (triangle), 126
  line, 126

D

differential operators, 127
dimensioning
  Ferrite inductance, 5
  Flyback power supply, 5, 72–74, 86, 100, 102–104
  Flyback transformer, 28, 32, 58, 88
  forward transformer, 23, 33, 88

F, G

formulas
  Euler, 112
  trigonometric, 112
Fresnel diagram, 111
Galvanic isolation voltage measurement, 97
Green-Ostrogradsky (theorem), 126

I, N

isolated
  forward converters, 19, 24, 33, 71
  power supplies, 7
  Flyback, 24
nabla, 128

P, R

phaser, 117
power
  active, 122, 123, 125
  apparent, 122, 123, 125

fluctuating, 122, 125
reactive, 123, 125
resonant inverter, 43, 46, 49, 51, 53, 57

S

snubbers, 35
soft switching, 41, 42
state modeling
continuous conduction, 2, 4, 5, 9, 13, 16, 21, 23, 60, 67
discontinuous conduction, 2, 5, 8, 10, 11, 13, 16–18, 25, 67, 68, 103, 106
Stokes-Ampère theorem, 127

T

three-phase system
balanced, 113, 114, 125
direct, 113
transfer functions, 59, 60, 62, 66, 67, 68, 100
transformation
Concordia, 115–118
Park, 116, 117

V, Z

value
average, 2, 4, 12–14, 28, 48, 49, 60, 103, 118, 119, 125
RMS, 29, 33, 74, 75, 82, 90, 112, 118, 119, 126
voltage
line-to-line, 125, 126
line-to-neutral, 125, 126
zero sequence component, 114, 115

Printed in the United States
By Bookmasters